Interesting Stories
for Unleashed Curiosity

50 SCIENTIFIC MARVELS FOR CURIOUS MINDS THAT WILL BLOW YOUR MIND

DEDICATION

To all the curious minds who look up at the stars and wonder, who ask "why" and "what if," and who believe that the world is full of marvels waiting to be discovered—this book is dedicated to you.

O. Bennet

Copyright © 2024 Ely & Oly books
All rights reserved.

INTERESTING STORIES
FOR UNLEASHED CURIOSITY
50 SCIENTIFIC MARVELS FOR CURIOUS MINDS THAT WILL BLOW YOUR MIND

These **50 Scientific Marvels** represent just a glimpse of the countless groundbreaking discoveries and innovations from the past decade. In this book, I've carefully selected those that have captivated my attention for their sheer ingenuity, the revolutionary technologies behind them, and their profound impact on our world.

From mind-bending advancements in quantum computing to the transformative potential of genetic editing, these stories showcase the incredible progress we've made, and they highlight the innovations that are reshaping our future.

O. Bennet

PROLOGUE

To the endlessly curious, the seekers of knowledge and those who marvel at the history of human progress: welcome. In the last decade, we have witnessed scientific and technological breakthroughs at a dizzying pace, pushing the boundaries of what we know and reshaping what we thought possible. This book is a tribute to that journey: a collection of what I believe to be the most interesting stories I have been able to compile that bring those wonders to life and capture the spirit of discovery that defines our time.

In these pages you will find stories ranging from advances in medicine that are transforming healthcare to feats of space exploration that are redefining our place in the cosmos, each story an invitation to marvel at what humanity can achieve when driven by curiosity and determination. These breakthroughs are not mere events; they are glimpses of a future that draws nearer with each revelation, and they speak to the power of human ingenuity to shape that future.

Each story invites you to experience that spark of wonder, to step into the lives of scientists, engineers, and visionaries who work tirelessly to push the boundaries of our knowledge. I hope you discover, as I did, that their passion is contagious, their ideas profound, and their accomplishments inspiring.
Thank you for joining this celebration of human curiosity. I hope and pray that this book will fuel your imagination, deepen your wonder, and remind you of the extraordinary potential we all harbor waiting to be awakened.

Let us embark on this journey together.

CONTENTS

001 Quantum Computing - The Dawn of Quantum Supremacy 006
002 CRISPR-Cas9 - Revolutionizing Genetic Engineering 009
003 mRNA Vaccines - The Fast-Tracked COVID-19 Solution 012
004 Gravitational Waves - Proving Einstein Right 015
005 First Image of a Black Hole - A New Window into the Universe 018
006 Self-Driving Cars - The Road to Autonomous Driving 021
007 Artificial Intelligence - The Age of Machine Learning and Deep Learning 024
008 Reusable Rockets - The Future of Space Travel 027
009 Fusion Energy - Getting Closer to Unlimited Clean Energy 030
010 Human Genome Editing - New Frontiers in Health 033
011 3D Printing - From Prototypes to Everyday Objects 036
012 Exoplanet Discoveries - Searching for Habitable Worlds 039
013 Graphene - The Wonder Material of the Future 042
014 Neuralink - Bridging the Brain-Technology Divide 045
015 Wearable Tech - Health Monitoring on Your Wrist 048
016 Solar Energy Breakthroughs - Powering the Future 051
017 AI in Art - Machines That Create Masterpieces 054
018 Blockchain and Cryptocurrencies - Revolutionizing Finance 057
019 Augmented Reality - Merging Real and Digital Worlds 061
020 Space Tourism - The Next Frontier for Travel 065
021 Advanced Prosthetics - Merging Biology with Technology 069
022 Plastic-Eating Enzymes - Tackling the Global Waste Problem 073
023 AI in Healthcare - Diagnosing Diseases with Precision 077
024 Quantum Cryptography - Securing the Future of Data 081
025 Hydrogen Fuel Cells - Clean Energy for Transportation 085
026 AI-Driven Drug Discovery - Finding New Cures Faster 089
027 Smart Homes - Living with Artificial Intelligence 093
028 Autonomous Drones - Revolutionizing Delivery and Transportation 097
029 Self-Healing Materials - The Next Generation of Manufacturing 101
030 Exoskeletons - Enhancing Human Mobility 105
031 Deep-Sea Exploration - Unlocking the Secrets of the Ocean 109
032 Advanced AI - Autonomous Warfare 113
033 Bioprinting - Printing Human Tissues and Organs 117
034 Carbon Nanotubes - The Strongest Material Known to Man 121
035 Next-Gen 5G Networks - Connecting the Futur 125
036 Smart Clothing - Technology Woven into Fabric 129
037 3D Printing Homes - Revolutionizing Construction 132
038 Space Mining - Extracting Resources from Asteroids 135
039 Human-Robot Collaboration - Merging the Strengths of Both 138
040 Digital Twins - Simulating the Real World in Digital Form 141
041 Synthetic Biology - Designing Life in the Lab 144

042 Discovery of Quantum Entanglement Applications - Instant Connections 147
043 Advances in Space Telescopes - Peering Deeper into the Cosmos 150
044 Solar Probe Missions - Touching the Sun 153
045 Dark Matter and Dark Energy Studies - Unveiling the Invisible Universe 156
046 Higgs Boson Found - The Missing Piece of the Particle Puzzle 159
047 Discovery of Liquid Water on Mars - Possibility of Life 162
048 Breakthrough Listen Project - Searching for Extraterrestrial Life 165
049 Discovery of Ancient Human Species - Redrawing Our Family Tree 168
050 Plant Communication Discoveries - The Secret Life of Plants 171
050+ Deepfake Technology - The New Era of Synthetic Media (just one more) 174

001 Quantum Computing
The Dawn of Quantum Supremacy

In October 2019, the world of computing experienced a seismic shift. Google announced that it had achieved **quantum supremacy**, a milestone many believed was still decades away. But what exactly does that mean, and why should we care? In simple terms, quantum supremacy refers to the moment when a **quantum computer** can solve a problem that even the world's most powerful classical supercomputers would take millennia to complete. The problem Google solved? A complex calculation using its **Sycamore** quantum processor, which took just 200 seconds. For perspective, the same calculation would have taken Summit, the fastest classical supercomputer, about 10,000 years.

But what makes **quantum computing** so revolutionary? To understand it, let's first talk about how traditional computers work. Your phone, laptop, or even the most advanced supercomputers use **bits** to process information. These bits are **binary**—meaning they can represent either a 0 or a 1. Every piece of digital information boils down to combinations of these **two states**.

Quantum computers, on the other hand, use **qubits**. Unlike classical bits, qubits can exist in both the 0 and 1 states simultaneously, thanks to a phenomenon known as **superposition**. Picture it like spinning a coin—while it's spinning, it's neither heads nor tails, but both. Furthermore, qubits can be **entangled**. This means that the state of one qubit is directly linked to the state of another, even if they are far apart, allowing **quantum computers** to process information in a way that defies our usual understanding of **physics**.

Google's experiment in 2019 took advantage of these principles to perform a task that classical computers would find exponentially difficult. They used

random circuit sampling, a test that involves generating random numbers through quantum circuits and sampling the output. While this particular task doesn't have practical applications in everyday life, it was a groundbreaking proof of concept—**quantum computers** could, indeed, outperform their classical counterparts for certain tasks.

This achievement marks a monumental step forward, but it's also just the beginning. **Quantum computers** are still in their infancy, and they face several challenges before they can be used in practical, everyday applications. For instance, **qubits** are highly sensitive to their environment. They require extremely low temperatures to maintain their quantum states, and even the slightest disturbance can cause errors, a phenomenon known as **decoherence**. Scientists are working on solutions like **quantum error correction** to make quantum computing more stable and scalable.

Applications:

While quantum computers aren't yet ready to be part of your daily life, their potential is vast and far-reaching. Here are a few key areas where quantum computing could revolutionize industries:

- **Cryptography:** One of the most anticipated applications is in the field of cryptography. Today's encryption methods rely on the difficulty of factoring large numbers, a task classical computers struggle with. A quantum computer, however, could crack these codes in seconds. But it's not all bad news—quantum computers could also be used to create unbreakable encryption through **quantum key distribution**, a method that leverages the laws of quantum mechanics to secure information in a way that's impossible to hack.

- **Drug Discovery:** Quantum computers could radically accelerate the process of drug discovery by **simulating** molecular interactions at a **quantum level**. This would allow scientists to predict how molecules behave, drastically reducing the time it takes to **design** and **test** new medications. Pharmaceutical companies are already exploring how quantum computing could help in the search for new treatments for diseases like **Alzheimer's** and **cancer**.

- **Materials Science:** Predicting the **properties** of new materials is another area where quantum computers could shine. By simulating the behavior of **atoms** and **molecules**, quantum computers can help scientists discover new materials with desirable properties—whether it's superconductors that conduct electricity without resistance or stronger, lighter materials for building and manufacturing.

Summary:
Google's achievement of **quantum supremacy** is more than just a headline—it's a pivotal moment in the history of computing. By proving that quantum computers can outperform classical ones, even for a highly specialized task, Google has opened the door to a new era of technological innovation. While we're still years away from seeing practical quantum computers in everyday use, the possibilities are staggering. From revolutionizing cryptography to speeding up drug discovery and unlocking new materials, **quantum computing** has the potential to reshape industries and solve problems that were previously thought unsolvable. As researchers overcome the challenges of error rates and scalability, the future of computing looks brighter—and stranger—than ever before.

002 CRISPR-Cas9
Revolutionizing Genetic Engineering

Imagine a world where genetic diseases could be cured with a simple edit, where crops could be tailored to thrive in any climate, and where we could eliminate pests without harmful chemicals. This isn't the plot of a science fiction movie; it's the promise of **CRISPR-Cas9**, one of the most revolutionary scientific breakthroughs of the last decade. In 2012, scientists Jennifer Doudna and Emmanuelle Charpentier discovered how to use CRISPR (Clustered Regularly Interspaced Short Palindromic Repeats) along with an enzyme called **Cas9** to edit DNA with unprecedented precision, efficiency, and ease.

CRISPR technology was first discovered as a **natural defense** mechanism used by **bacteria**. In nature, bacteria use CRISPR to remember and defend against **viruses** by cutting their **DNA**. Scientists realized they could harness this system as a gene-editing tool, effectively creating a biological **"scissors"** capable of cutting and modifying DNA at specific locations. The implications were enormous: we could now alter the genetic code—the **blueprint of life** itself—with remarkable accuracy.

Before CRISPR, gene editing was a laborious, expensive, and time-consuming process. Technologies like **Zinc Finger Nucleases** and **TALENs** existed but were far less efficient and far more complex. CRISPR, on the other hand, is relatively simple. It uses a small guide RNA to direct the **Cas9 enzyme** to a specific sequence of DNA, where it makes a cut. Once the DNA is cut, the cell's natural repair mechanisms kick in, either disabling the gene or allowing scientists to insert a new one. This breakthrough made gene editing accessible to a wider range of researchers, sparking a revolution in biology and medicine. The potential of CRISPR seems almost limitless.

From curing genetic diseases like **sickle cell anemia** and **muscular dystrophy** to engineering crops that are resistant to pests, CRISPR opens doors to countless possibilities. But with such incredible power comes equally significant ethical challenges.

Editing the genes of embryos—**germline editing**—is particularly controversial because the changes would be passed down to future generations. This raises the specter of **designer babies**, where parents might one day select traits like intelligence, height, or athleticism for their children. While the scientific community has largely agreed that CRISPR should be used for medical purposes, the ethical debate is ongoing.

One of the most publicized examples of CRISPR's power came in 2018, when a Chinese scientist claimed to have edited the genes of twin babies to make them resistant to **HIV**. This announcement sent shockwaves through the scientific community, sparking an international outcry and calls for stricter **regulations** on human gene editing. While the potential for medical breakthroughs is immense, the need for **ethical guidelines** and oversight is critical.

Applications:

The versatility of CRISPR-Cas9 has led to groundbreaking applications in medicine, agriculture, and even environmental conservation:

- **Medicine:** CRISPR has already been used in clinical trials to treat **sickle cell anemia** and **beta-thalassemia** by editing the faulty genes responsible for these conditions. Scientists are also working on using CRISPR to treat more complex diseases, such as **cancer**, by editing the genes of immune cells to target tumors more effectively.

- **Agriculture:** CRISPR is being used to engineer crops that are more resistant to pests, diseases, and environmental stresses like drought. This has the potential to **improve** food security and reduce the need for harmful pesticides. For example, scientists have edited rice to be resistant to a common fungus that devastates crops, potentially **saving** millions of tons of food each year.

- **Environmental Conservation:** Researchers are exploring the use of CRISPR to control invasive species and protect endangered ones. For instance, CRISPR has been used to modify mosquitoes in a way that could reduce the spread of diseases like **malaria** by making them unable to transmit the parasite. Similarly, scientists are considering using gene editing to bring back extinct species, like the woolly mammoth, through a technique known as **de-extinction**.

Summary:

CRISPR-Cas9 has revolutionized the field of **genetic engineering**, offering scientists a powerful tool to edit **DNA** with unprecedented precision. Its applications range from curing genetic diseases and revolutionizing agriculture to possibly eliminating pests and even resurrecting extinct species. However, with this power comes significant **ethical concerns**, particularly regarding human gene editing and its long-term implications. As the science of gene editing progresses, so too must the conversation about how to use this technology **responsibly**. Nevertheless, CRISPR stands as one of the most promising and transformative breakthroughs of the 21st century.

003 mRNA Vaccines
The Fast-Tracked COVID-19 Solution

The COVID-19 pandemic swept across the globe in 2020, changing life as we knew it. As the virus spread, the world turned its attention to finding a solution, and fast. That solution came in the form of **mRNA vaccines**, a technology that was rapidly developed and deployed, saving millions of lives. But how did these vaccines, which were previously untested on such a massive scale, manage to halt the biggest pandemic in a century?

Messenger RNA **(mRNA)** vaccines work differently from **traditional** vaccines, which usually introduce weakened or inactivated viruses to trigger an immune response. Instead, **mRNA** vaccines **instruct cells** in the body to produce a **protein**—specifically, the spike protein found on the surface of the **SARS-CoV-2** virus. Once this protein is produced, the immune system recognizes it as foreign and begins generating antibodies, teaching the body how to fight the virus without ever being exposed to the actual pathogen.

This method of vaccine development was **revolutionary**, not just in its effectiveness, but in how quickly it could be produced. Traditional vaccines can take years to develop, test, and manufacture, but **mRNA vaccines** bypass many of those steps. Once the genetic sequence of a virus is known, scientists can quickly design an **mRNA vaccine** to target it. This is exactly what happened with the COVID-19 vaccines

produced by Pfizer-BioNTech and Moderna, which were among the first to be authorized for emergency use in late 2020.

What's even more impressive is that while **mRNA vaccines** were fast-tracked for COVID-19, the technology behind them had been in development for decades. Researchers had been studying **mRNA** as a potential vaccine platform for years, hoping to use it to fight diseases like **Zika**, **rabies**, and **influenza**. The COVID-19 pandemic simply accelerated the application of this already promising technology, bringing it to the forefront of medical innovation.

Despite the rapid timeline, mRNA vaccines proved to be not only **safe** but **highly effective**, with trials showing around 95% efficacy in preventing symptomatic COVID-19 infections. The speed at which they were developed and approved was **unprecedented**, but so was the situation. Scientists worked around the clock, and governments fast-tracked approvals while maintaining **rigorous safety standards**. The results speak for themselves: by the end of 2021, over 1 billion doses of mRNA vaccines had been administered, helping to reduce the severity and spread of COVID-19 **globally**.

Applications:

The successful deployment of mRNA vaccines for COVID-19 opens up a world of possibilities for other diseases:

- **Infectious Diseases:** Now that **mRNA technology** has been validated, researchers are turning their attention to other infectious diseases. Influenza, which still causes significant illness and death annually, is one target. Scientists are developing **mRNA flu vaccines** that could be more effective and faster to update than current vaccines.

- **Cancer:** The potential for **mRNA vaccines** goes beyond infectious diseases. Researchers are exploring how this technology could be used to create **cancer vaccines**. These vaccines would instruct the immune system to recognize and attack cancer cells, offering a new form of immunotherapy for treating the disease.

- **Zoonotic Diseases:** Diseases that jump from animals to humans, such as **Zika**, **Ebola**, and **avian flu**, could be tackled more swiftly with **mRNA vaccines**. The ability to quickly design a vaccine once the genetic sequence of a new virus is known could be key to stopping future pandemics before they spiral out of control.

Summary:

The development of **mRNA vaccines** for **COVID-19** was a groundbreaking achievement in medicine, allowing scientists to create, test, and deploy a highly effective vaccine in record time. This innovative approach to vaccination offers hope for faster, more effective vaccines for other infectious diseases and even cancer in the future. The success of the **mRNA COVID-19 vaccines** marks the beginning of a new era in medicine, one where genetic information can be harnessed to protect and save lives more quickly than ever before. As the technology continues to evolve, the world will be better prepared for future pandemics and health challenges.

004 Gravitational Waves
Proving Einstein Right

In 2015, a century-old prediction by **Albert Einstein** was finally confirmed: **gravitational waves** exist. The groundbreaking discovery shook the scientific community and opened a new window into the universe. But what exactly are gravitational waves, and why was this such a significant moment for physics?

Gravitational waves are ripples in the fabric of **spacetime** caused by the acceleration of massive objects, such as merging black holes or neutron stars. Einstein first predicted their existence in 1915 as part of his **General Theory of Relativity**. According to Einstein's theory, when massive objects move, they disturb spacetime, much like a stone creates ripples when thrown into a pond. These ripples, or gravitational waves, travel through the universe at the speed of light, carrying information about the cataclysmic events that created them.

For decades, scientists tried to detect these elusive waves, but it wasn't until September 14, 2015, that the **Laser Interferometer Gravitational-Wave Observatory (LIGO)** made the first direct observation. The signal came from the collision and merger of two black holes located over a billion light-years away, a cosmic dance that unleashed enormous energy. The detected wave lasted only a fraction of a second but marked a monumental moment in scientific history. It was the first direct evidence of gravitational waves and the first time scientists had observed a **black hole** merger. The discovery confirmed Einstein's theory in a way that was once purely theoretical.

The technology behind LIGO is nothing short of incredible. LIGO consists of two huge **interferometers**, one in **Washington State** and the other in

Louisiana, each with two 4-kilometer-long arms arranged in an L-shape. These arms use laser beams to measure tiny changes in distance caused by passing gravitational waves—changes so minuscule that they are smaller than a fraction of a proton's width. When a gravitational wave passes through Earth, it slightly stretches and squeezes spacetime, and LIGO's sensitive detectors are able to pick up on this disturbance.

This discovery was significant not only because it **confirmed** Einstein's theory, but also because it gave scientists a new way to study the universe. Before LIGO, our understanding of the **cosmos** was based almost entirely on observing **electromagnetic radiation**, such as light. Now, with **gravitational waves**, we have a **new "sense"** that allows us to detect and study events that were previously invisible. It's like being able to both see and hear the universe for the first time.

Applications:

The detection of gravitational waves has opened up new avenues of research and discovery in physics and astronomy:

- **Studying Black Holes:** Gravitational waves provide us with the best way to observe **black hole mergers**. Black holes don't emit light, so traditional telescopes cannot detect them. But when two black holes collide, they generate powerful gravitational waves that can now be detected, allowing scientists to study their properties and behavior in ways that were impossible before.

- **Understanding Neutron Stars:** In 2017, LIGO and its European counterpart, **Virgo**, detected gravitational waves from the **collision of two neutron stars**—another first. This discovery provided scientists with insights into the structure of neutron stars, the densest objects in the universe, and helped to explain the origins of heavy elements like gold and platinum, which are formed during such collisions.

- **Mapping the Universe:** By detecting gravitational waves from various sources, scientists can create a new map of the universe. These waves carry information about the objects that created them, giving us clues about the nature of **dark matter**, the formation of galaxies, and even the expansion rate of the universe.

Summary:
The detection of **gravitational waves** was a monumental breakthrough, confirming a key aspect of **Einstein's General Theory of Relativity** and providing scientists with a new tool to explore the universe. By "listening" to the universe through gravitational waves, we can now study cosmic events that were previously undetectable, such as **black hole mergers** and **neutron star collisions**. This discovery has opened up new opportunities for understanding the most mysterious and powerful forces in the cosmos, marking the dawn of **gravitational wave astronomy**. With each new detection, we learn more about the hidden workings of the universe, proving that there is still so much left to discover.

005 First Image of a Black Hole
A New Window into the Universe

For centuries, black holes were one of the most mysterious and intriguing phenomena in the cosmos. These regions of spacetime, **where gravity is so strong that not even light can escape**, have fascinated scientists and the public alike. Although theorized by Albert Einstein in 1915 as part of his General Theory of Relativity, no one had ever directly observed a black hole—until 2019, when a global team of scientists captured the **first-ever image of a black hole**, a feat that marked a monumental leap forward in astronomy and physics.

The black hole in question resides in the center of the **Messier 87** (M87) **galaxy**, about 55 million light-years away from Earth. It's **a supermassive black hole**, with a mass 6.5 billion times greater than our Sun. The image was captured by the **Event Horizon Telescope (EHT)**, a network of radio telescopes spanning the globe. By synchronizing their efforts, these telescopes essentially created a planet-sized observational tool, capable of capturing an image of a region of space once thought to be invisible.

The iconic image shows a **dark circular silhouette**, the so-called **event horizon**, surrounded by a glowing **ring** of **light**. This light comes from the superheated gas and dust swirling around the black hole at nearly the **speed of light** before being pulled into the **abyss**. The image **confirmed** decades of theoretical predictions and provided direct visual proof of the **existence** of black holes. What had once been considered a theoretical curiosity was now an observable **reality**.

Capturing this image required a level of precision and ingenuity that had never before been achieved in astronomy. The EHT team used a technique called **very long baseline interferometry (VLBI)**, which allowed multiple telescopes across different continents to work together as if they were one enormous telescope. The data collected from each telescope was synchronized using **atomic clocks** and combined by a series of **supercomputers**. This collaboration involved over **200 researchers** from more than **20 countries**, all working toward the same goal: to take the **first picture** of a black hole.

The significance of this image cannot be overstated. It gave scientists a new way to test the **predictions** of **Einstein's General Theory of Relativity** under the most extreme conditions. The shape and size of the shadow matched the theoretical predictions based on Einstein's equations, further validating one of the most important scientific theories of all time. But beyond that, the image represented humanity's first glimpse of one of the **universe's most enigmatic objects**—a glimpse that sparked both awe and curiosity.

Applications:

The successful imaging of a black hole has far-reaching implications for astrophysics and could lead to new discoveries in various fields:

- **Testing General Relativity:** The first black hole image provides a unique environment to test the predictions of **Einstein's General Theory of Relativity** in the strong-gravity regime. While relativity has been tested in weaker gravitational fields, such as those near planets and stars, black holes offer the most extreme conditions. Future observations could reveal subtle deviations that might point to new physics beyond Einstein's theory.
- **Understanding Supermassive Black Holes:** Observing the black hole at the center of **M87** gives scientists valuable data to study the behavior of **supermassive black holes**. These enormous objects are thought to reside at the center of most galaxies, including our own **Milky Way**, and play a crucial role in galaxy formation and evolution. The data from the EHT could help explain how these black holes interact with their host galaxies and how they grow over time.
- **Advancing Telescope Technology:** The methods used to capture the first image of a black hole, particularly **VLBI**, have pushed the boundaries of what is possible in observational astronomy. The EHT's success has already spurred plans to build even more sensitive telescopes that could

image black holes with greater detail, allowing scientists to observe **accretion disks, jets,** and other phenomena that occur near black holes. This could deepen our understanding of how black holes affect their surroundings.

Summary:

In 2019, the world saw the **first-ever image of a black hole,** providing direct visual evidence of one of the most mysterious objects in the universe. Captured by the **Event Horizon Telescope,** this achievement validated predictions from **Einstein's General Theory of Relativity** and gave scientists a new tool to study black holes. The image of the **black hole** in **M87** offers a unique opportunity to test theories of gravity in extreme environments, advance our understanding of supermassive black holes, and improve our observational capabilities in astronomy. This historic moment opened a new window into the universe, proving that even the darkest corners of space can reveal their secrets when viewed with the right tools.

006 Self-Driving Cars
The Road to Autonomous Driving

For over a century, the automobile has been a symbol of personal freedom, allowing people to move from one place to another at will. But in recent years, the advent of **self-driving** cars has begun to change what it means to travel. These vehicles, once the stuff of science fiction, are now on the verge of becoming an everyday reality, thanks to advancements in **artificial intelligence (AI)**, sensors, and computing power. The dream of autonomous driving, where cars can navigate streets and highways without human intervention, is closer than ever.

The road to self-driving cars started decades ago, with incremental improvements in **driver-assistance systems** like cruise control, anti-lock brakes, and parking sensors. But the real breakthrough came with the development of **artificial intelligence** and **machine learning**, which allowed cars to "see" and "think" in real-time. Companies like Tesla, Waymo (a subsidiary of Google), and Uber have invested billions of dollars into developing autonomous driving technologies. These systems use a combination of **LIDAR (Light Detection and Ranging), radar, cameras**, and **GPS** to create a detailed map of the surrounding environment and navigate it safely.

The technology behind self-driving cars is impressive. LIDAR systems send out **laser pulses** that bounce off objects and return to the sensor, creating a precise, **3D map** of the environment. Cameras and radar provide additional information, allowing the car to detect things like traffic lights, road signs, and pedestrians. The vehicle's AI processes this data and makes **split-second decisions**, such as when to stop, accelerate, or turn, all while avoiding collisions. The more advanced systems, like those being developed by Waymo, are designed to operate without any human input, eliminating the need for steering wheels or pedals.

However, the road to full autonomy has been anything but smooth. One of the biggest challenges facing self-driving cars is the **complexity of real-world driving environments**. While autonomous vehicles perform well in controlled environments like highways, urban areas present a different set of challenges. Navigating through city streets filled with pedestrians, cyclists, and unpredictable drivers requires an **enormous amount of data processing**. Moreover, the vehicle's AI must learn to handle rare but critical situations, such as sudden obstacles or changing weather conditions, all while making decisions that prioritize safety.

Another challenge has been public trust and **regulatory** approval. The idea of sitting in a car that **drives itself** is exciting for some but terrifying for others. High-profile accidents involving self-driving cars have raised **concerns** about safety, even though human error accounts for the vast majority of traffic accidents today. Governments and regulatory bodies are still figuring out how to create **laws** and **infrastructure** to support self-driving cars, and widespread adoption may depend as much on legal and social issues as on technological advances.

Applications:

While fully autonomous vehicles are not yet commonplace, self-driving technology is already being applied in several ways, with more advancements on the horizon:

- **Ridesharing and Delivery Services:** Companies like Waymo and Uber are **testing** self-driving vehicles for ridesharing and delivery services. The idea is that autonomous cars could operate around the clock without the need for a human driver, making transportation more efficient and reducing the cost of services. Amazon is also exploring autonomous delivery **trucks** and **drones** to revolutionize logistics.

- **Long-Distance Freight:** Autonomous technology is set to have a significant impact on the trucking industry. Self-driving trucks are already being tested to transport goods across long distances. These vehicles can **drive continuously** without rest, improving delivery times and reducing human error, which is a major cause of accidents. Companies like TuSimple and Embark are leading the charge in this area.

- **Improved Traffic Flow and Safety:** One of the promises of self-driving cars is the potential to **reduce accidents** and **improve traffic efficiency**. Autonomous cars can communicate with one another, reducing the chances of accidents caused by human error. They can also optimize traffic flow by avoiding **congestion**, leading to less time spent in traffic and **fewer emissions**. This would have a massive impact on urban planning and sustainability.

Summary:

Self-driving cars represent one of the most significant technological advancements of the 21st century. Through the integration of **AI, LIDAR, radar**, and **machine learning**, vehicles are now capable of navigating complex environments **without human intervention**. While the technology is still evolving and faces hurdles in terms of safety, regulation, and public trust, the potential applications are vast—from ridesharing and delivery services to revolutionizing freight transportation and improving traffic flow. As autonomous driving becomes more reliable and widespread, it could drastically reshape our cities, reduce accidents, and redefine the way we think about mobility. The road to full autonomy is long, but **self-driving cars** are paving the way for a future of safer, smarter transportation.

007 Artificial Intelligence
The Age of Machine Learning and Deep Learning

In recent years, **artificial intelligence (AI)** has evolved from a futuristic concept to a transformative force reshaping industries, businesses, and daily life. While the idea of creating **machines** that can **"think"** has been around for decades, the real revolution began with the advent of **machine learning** and **deep learning**—technologies that enable computers to learn from vast amounts of data and make decisions or predictions without being explicitly programmed for every task. We are now living in the age of AI, where algorithms are capable of performing tasks once thought to be the sole domain of humans.

Machine learning is a subset of AI that focuses on **enabling computers to learn from data**. Traditional software relies on predefined rules and logic to perform tasks, but machine learning algorithms analyze data, identify patterns, and improve over time. The more data the system processes, the better it becomes at making predictions. For example, Netflix uses machine learning to recommend shows and movies based on user preferences, while Google relies on it for search results and language translation.

Deep learning, a more advanced form of machine learning, has taken this concept even further by mimicking the neural networks of the human brain. These **artificial neural networks** consist of multiple layers of **algorithms** that process information in a hierarchical manner. Deep learning models excel at tasks like image recognition, natural language processing, and even generating human-like text. One of the most well-known examples is OpenAI's GPT-3, a deep learning model capable of writing coherent articles, stories, and even answering complex questions in natural language.

The rise of AI can largely be attributed to the explosion of **big data** and advances in **computing power**. With the advent of cloud computing and more powerful hardware, it became possible to process massive datasets quickly and efficiently. This data feeds AI models, allowing them to learn and make increasingly accurate predictions. At the same time, advancements in **graphics processing units (GPUs)** and specialized hardware have accelerated the training of deep learning models, making AI more accessible and practical for everyday use.

One of the key reasons AI has become so prominent is its **ability** to automate tasks at scale. From fraud detection in banking to predictive maintenance in manufacturing, AI can **analyze massive** datasets in **real-time**, identify patterns, and recommend or execute decisions. In healthcare, AI models can scan medical images to detect conditions like cancer more accurately than human doctors. In retail, **AI-powered chatbots** handle customer service inquiries and assist in online shopping.

However, the rise of AI hasn't been without **challenges** and **controversies**. **Ethical concerns** surrounding AI have sparked heated debates. One of the primary concerns is **bias in AI systems**, where algorithms trained on biased data may make unfair decisions, especially in sensitive areas like hiring, lending, and criminal justice. Additionally, the rapid advancement of AI has raised questions about **job displacement**, as machines and algorithms take over tasks once performed by humans. AI's growing influence also stirs concerns about privacy and surveillance, particularly as facial recognition technology becomes more widespread.

Applications:

AI, and specifically machine learning and deep learning, have found applications across nearly every industry. Here are a few key examples of how these technologies are reshaping the world:

- **Healthcare:** AI-powered systems are revolutionizing healthcare by improving diagnostics, predicting patient outcomes, and personalizing treatments. Deep learning models can analyze medical images to detect conditions like **lung cancer** and **heart disease** more accurately and faster than traditional methods. AI-driven drug discovery is also accelerating the development of **new medications** by predicting how different compounds interact with the human body.

- **Autonomous Vehicles:** Self-driving cars rely heavily on deep learning

algorithms to navigate roads, avoid obstacles, and make **split-second decisions**. Companies like Tesla and Waymo are at the forefront of using AI to develop vehicles that can operate without human input, potentially reducing traffic accidents caused by human error.

- **Natural Language Processing (NLP):** AI systems that **understand** and generate **human language** are becoming increasingly common. Virtual assistants like Siri and Alexa use NLP to understand **voice commands**, while deep learning models like GPT-3 are capable of generating human-like text. These technologies are used for **chatbots**, **translation services**, and even **writing software**.

Summary:

We are living in the age of **artificial intelligence**, where **machine learning** and **deep learning** technologies are transforming industries and daily life. AI-powered systems are automating tasks at unprecedented scales, from diagnosing diseases and powering self-driving cars to generating human-like language. While the potential of AI is immense, it also comes with ethical concerns, including bias, privacy, and job displacement. As we move deeper into the era of AI, the challenge will be to harness its power responsibly while navigating the societal changes it brings. Nevertheless, the age of AI has only just begun, and its impact on the world will continue to grow.

008 Reusable Rockets
The Future of Space Travel

For decades, space travel was synonymous with **immense costs** and **waste**. Rockets that launched into space were **single-use machines**, discarded after delivering their payloads, rendering space exploration an expensive and unsustainable endeavor. Enter SpaceX, the private aerospace company founded by Elon Musk, which revolutionized the industry with the development of **reusable rockets**. This breakthrough has transformed space travel, making it more affordable, sustainable, and frequent—paving the way for future missions to the Moon, Mars, and beyond.

The core problem that SpaceX set out to solve was the **throwaway nature** of rockets. In traditional space missions, rockets were designed for **one-time use**. After propelling their payload—whether it be satellites, scientific instruments, or astronauts—into orbit, most of the rocket components were jettisoned and left to fall back to Earth or float in space. This **"disposable"** approach made each launch extraordinarily costly, as new rockets had to be built for every mission.

In 2015, SpaceX achieved what many had thought impossible: they successfully landed the **Falcon 9** rocket's first stage back on Earth after delivering a payload to space. This stage is the most expensive part of the rocket, responsible for propelling the vehicle out of Earth's atmosphere. By **recovering** and **reusing** this part, SpaceX slashed the costs of launching rockets, making space travel significantly cheaper. This achievement was the culmination of years of research and development, and it marked a turning point in the history of space exploration.

SpaceX's **reusable rocket** technology is centered around two key innovations. First, the Falcon 9 rocket is equipped with **landing legs** and **small thrusters** that allow it to perform a controlled descent after completing its mission. Second, it uses **grid fins** to stabilize itself as it reenters the atmosphere. Once the rocket reaches a certain altitude, its engines reignite to slow its descent and guide it toward a precise landing on either a droneship in the ocean or a landing pad on solid ground. The process is so refined that SpaceX has successfully landed and reused rockets multiple times, significantly reducing the cost per launch.

Reusable rockets aren't just about cost savings—they represent a **new era of sustainability in space exploration.** In a world where sustainability is becoming increasingly important, reusing rockets aligns with global efforts to reduce waste and improve resource efficiency. Furthermore, the ability to reuse rockets means that space travel can happen more **frequently**, as there's no need to build new rockets for each mission. This opens up the possibility for more regular and ambitious **space missions**, including those that aim to establish a human presence on Mars.

SpaceX has continued to push the boundaries of reusable rocket technology. The company's Falcon Heavy, the most powerful operational rocket in the world, is also partially reusable. And the next big leap? Starship, a **fully reusable spacecraft** that SpaceX hopes will one day carry humans to Mars. Unlike the Falcon 9, Starship is designed for complete reusability, with both its first and second stages capable of landing and being launched again. This innovation could drastically reduce the costs of **deep space exploration** and make **interplanetary travel** a reality.

Applications:

The development of reusable rockets by SpaceX has far-reaching implications for space exploration, satellite deployment, and even future human space travel:

- **Lowering the Cost of Space Missions:** By reusing rockets, SpaceX has dramatically reduced the cost of launching payloads into space. This reduction in cost makes space more accessible not only to governments but also to private companies. For example, the cost of launching a payload on a Falcon 9 rocket is estimated to be around $62 million, significantly lower than traditional rockets. This affordability is attracting more companies interested in launching satellites for telecommunications, Earth observation, and scientific research.

- **Enabling Ambitious Space Missions:** With the cost of space travel decreasing, more frequent and ambitious missions become possible. SpaceX's reusable rockets are critical to plans for returning humans to the **Moon** through **NASA's Artemis program**, and they are also central to Musk's vision of establishing a human settlement on **Mars**. Reusable rockets reduce the financial burden of deep space exploration, enabling sustained missions to other planets.

- **Commercial Space Travel:** The success of reusable rockets also opens the door to **space tourism**. SpaceX's collaboration with companies like Axiom Space aims to take private citizens to the **International Space Station (ISS)**. In the future, as reusability becomes more advanced, it's likely that space tourism could expand beyond low-Earth orbit, offering private individuals the chance to visit the Moon or even Mars.

Summary:
SpaceX's development of **reusable rockets** represents a paradigm shift in the world of space exploration. By recovering and reusing rocket stages, SpaceX has drastically reduced the cost of launching payloads into space, making frequent missions more affordable and sustainable. This breakthrough has not only revolutionized satellite launches but also paved the way for ambitious space endeavors, including human missions to the **Moon** and **Mars**. The ability to reuse rockets could also bring the dream of **space tourism** closer to reality. As the technology continues to evolve, reusable rockets are transforming **space travel**, making it accessible, cost-effective, and sustainable in ways that were once unimaginable.

009 Fusion Energy
Getting Closer to Unlimited Clean Energy

Imagine a world where we have access to nearly **limitless energy**, without the harmful emissions of fossil fuels or the dangerous byproducts of nuclear fission. This is the promise of **fusion energy**, a technology that has long been the holy grail of clean energy. While we're still years away from achieving commercially viable fusion power, recent breakthroughs have brought us closer than ever to harnessing the same process that powers the **Sun** and stars—nuclear fusion—right here on Earth.

At its core, **fusion energy** is the process of fusing two lighter atomic nuclei to form a heavier nucleus, releasing an immense amount of energy in the process. Unlike **nuclear fission**, which powers today's nuclear reactors by splitting atoms apart, fusion joins them together. The fuel for fusion is typically **hydrogen isotopes** like **deuterium** and **tritium**, which are abundant on Earth and generate little to no harmful byproducts. The end result? A clean, nearly inexhaustible energy source that could revolutionize how we power our world.

The process of **nuclear fusion** is incredibly challenging to replicate on Earth because it requires extreme conditions—**temperatures** of over **100 million degrees Celsius** and immense **pressure**—to overcome the natural repulsion between atomic nuclei. This is why fusion has **remained elusive** despite decades of research. However, significant advancements in the past decade have shown that we are getting closer to unlocking this powerful source of **energy**.

One of the most promising fusion projects is the **International Thermonuclear Experimental Reactor (ITER)**, a multinational collaboration involving 35 countries. Located in the south of France, ITER is designed to be the world's largest **tokamak**—a doughnut-shaped device that uses powerful **magnetic**

fields to confine and control the superheated **plasma** where fusion occurs. Once operational, ITER aims to produce 10 times the amount of energy it consumes, making it a critical step toward **commercial** fusion reactors. If successful, it could demonstrate that fusion energy is not only possible but scalable.

Another major player in the race to develop fusion energy is **Commonwealth Fusion Systems (CFS)**, a private company that spun out of research conducted at **MIT**. CFS is working on developing **high-temperature superconducting magnets**, which could drastically reduce the size and cost of fusion reactors. These magnets allow for stronger magnetic fields in a smaller space, making it easier to control the plasma and sustain the fusion reaction. In 2021, CFS achieved a breakthrough by successfully demonstrating a powerful magnetic field using these superconductors, bringing the dream of compact, affordable fusion reactors one step closer to reality.

But it's not just about building bigger and better reactors. Scientists are also exploring alternative approaches to fusion, such as **inertial confinement fusion (ICF)**, which uses powerful lasers to compress and heat tiny fuel pellets until they ignite and fuse. The **National Ignition Facility (NIF)** in the United States has been leading the charge in this area, achieving significant milestones in creating the extreme conditions necessary for fusion.

Applications:

If harnessed successfully, fusion energy could revolutionize the way we generate power and bring about a new era of clean energy. Here's how fusion could impact the world:

- **Limitless Clean Energy:** Fusion offers a nearly limitless supply of energy because its primary fuel sources, deuterium and tritium, are abundant on Earth. Deuterium can be extracted from seawater, while tritium can be produced from lithium, a common element. This means that, unlike fossil fuels, we won't run out of fuel for fusion anytime soon, making it a truly sustainable energy source.

- **Zero Greenhouse Gas Emissions:** Fusion doesn't produce greenhouse gases, meaning it won't contribute to global warming or climate change. The only byproduct of the fusion process is helium, an inert and harmless gas. If fusion reactors become commercially viable, they could play a crucial role in reducing our dependence on fossil fuels and mitigating the impacts of climate change.

- **Minimal Nuclear Waste:** Unlike nuclear fission, which produces long-lived radioactive waste, the waste products from fusion are far less hazardous. Tritium, one of the fuels used in fusion, is slightly radioactive, but its half-life is short, and it poses much less risk compared to the waste from fission reactors. Moreover, fusion reactions don't carry the risk of a nuclear meltdown, making fusion reactors much safer than traditional nuclear power plants.

Summary:
Fusion energy holds the promise of **unlimited, clean energy** that could power our world without contributing to climate change or producing hazardous waste. Recent breakthroughs in magnetic confinement, such as those at **ITER** and Commonwealth Fusion Systems, as well as advances in inertial confinement fusion at the National Ignition Facility, have brought us closer than ever to achieving **fusion power**. While there are still significant challenges ahead, the progress made in recent years suggests that fusion could one day become a reality, transforming the way we generate electricity and ushering in a **new era** of sustainable energy.

010 Human Genome Editing
New Frontiers in Health

The ability to edit the **human genome** has long been a dream for scientists and medical professionals alike. The potential to correct genetic disorders, eliminate hereditary diseases, and even enhance human traits has captivated the scientific community for decades. In recent years, thanks to groundbreaking advances in technologies like **CRISPR-Cas9, genome editing** is no longer science fiction—it is a rapidly developing field poised to revolutionize healthcare. With these tools, scientists can now make precise changes to the DNA of living organisms, opening up new frontiers in medicine and genetics.

At the heart of this revolution is **CRISPR-Cas9**, a gene-editing technology derived from a natural defense mechanism used by bacteria to fend off viruses. In essence, CRISPR allows scientists to target a specific sequence in the DNA, cut it with the **Cas9 enzyme**, and either disable a gene or insert new genetic material. This level of precision was unheard of before CRISPR's development, and it has sparked a wave of research into how we can use genome editing to treat, and even cure, a wide range of genetic disorders.

The promise of genome editing lies in its ability to treat diseases that were previously thought to be incurable. For example, **sickle cell anemia** is a devastating genetic condition that affects millions of people worldwide. The disease is caused by a mutation in the **HBB gene**, which affects the production of hemoglobin in red blood cells. Traditional treatments for sickle cell anemia involve managing symptoms and preventing complications, but CRISPR offers a more **radical solution**: directly correcting the faulty gene. In 2020, doctors successfully used CRISPR to treat a patient with sickle cell anemia by editing the patient's bone marrow cells to produce healthy red blood cells. This was a **landmark moment** in medical history, proving that genome editing could cure diseases at the genetic level.

However, genome editing's potential extends far beyond sickle cell anemia. Other genetic disorders, such as **cystic fibrosis, muscular dystrophy**, and certain forms of **cancer**, could one day be treated or even eradicated using CRISPR. In the case of cancer, for instance, CRISPR could be used to edit immune cells to better **recognize** and **destroy** cancer cells, offering a more personalized and targeted form of treatment. Researchers are also exploring the possibility of using genome editing to combat **HIV** by modifying immune cells to be resistant to the virus.

Yet, the power of genome editing also brings with it significant ethical challenges. The ability to edit the human genome raises questions about where we should draw the line. While most agree that using CRISPR to cure debilitating diseases is a positive development, the possibility of **germline editing**—modifying the **DNA** of embryos in a way that would pass changes on to future generations—has sparked intense debate. Germline editing could, in theory, eliminate genetic diseases before a child is born, but it also opens the door to the controversial idea of **designer babies**, where parents could choose their child's physical and intellectual traits. The global scientific community has called for strict **regulation** of germline editing, with many urging that it only be used for treating severe genetic disorders and not for enhancing human traits.

Applications:

The ability to edit the human genome has the potential to revolutionize medicine and healthcare. Here are some of the most promising applications of genome editing:

- **Treating Genetic Disorders: CRISPR-Cas9** is already being used in clinical trials to treat a variety of genetic disorders. In addition to the landmark case of sickle cell anemia, researchers are exploring treatments for conditions like **cystic fibrosis**, which is caused by a mutation in the **CFTR gene**. By correcting this mutation, CRISPR could offer a permanent cure for the disease rather than just managing its symptoms.

- **Cancer Immunotherapy:** Genome editing holds immense promise for improving **immunotherapy**, a type of cancer treatment that harnesses the patient's own immune system to fight the disease. By editing **T-cells** (a type of immune cell) to better recognize and attack cancer cells, scientists are developing personalized therapies that could be more effective and less harmful than traditional treatments like chemotherapy and radiation.

- **Infectious Disease:** Genome editing is also being explored as a treatment for infectious diseases like **HIV**. By editing the genes of immune cells, researchers aim to create cells that are resistant to the virus, potentially leading to a functional cure for HIV. This approach could one day be used to combat other viruses and pathogens, revolutionizing how we treat infectious diseases.

Summary:
Human genome editing is unlocking new possibilities in healthcare, offering hope for the treatment and even eradication of genetic diseases like **sickle cell anemia** and **cystic fibrosis**. Technologies like **CRISPR-Cas9** allow scientists to make precise changes to the DNA, potentially curing diseases at their genetic root. However, the power of genome editing also raises significant ethical questions, particularly regarding **germline editing** and its potential impact on future generations. As this technology continues to evolve, it will be crucial to establish guidelines that balance its immense potential with responsible and ethical use. The future of medicine is at the crossroads of genetics, and genome editing is leading the way into a new era of personalized, targeted healthcare.

011 3D Printing
From Prototypes to Everyday Objects

When **3D printing** first emerged, it was hailed as a game-changing technology for manufacturing and prototyping. Initially, it was used primarily by engineers and designers to create models and test ideas. However, over the past decade, **3D printing** has evolved far beyond its early applications. It has moved from the realm of rapid prototyping to producing functional, everyday objects, transforming industries like medicine, automotive, aerospace, and even fashion. Today, 3D printing is more than just a tool for innovation—it's a technology that's reshaping how we design, manufacture, and consume products.

At its core, 3D printing, also known as **additive manufacturing**, is a process that creates three-dimensional objects layer by layer using digital files. Traditional manufacturing typically involves **subtractive processes**, where material is removed from a larger block to shape a product. In contrast, **additive manufacturing** builds objects from the ground up, minimizing waste and enabling more complex designs. The materials used in 3D printing can range from **plastics** and **resins** to **metals** and even **biological materials** for medical applications.

One of the most significant advancements in 3D printing over the past decade has been its increased accessibility. Once limited to large companies and research institutions, **3D printers** are now available to hobbyists, schools, and small businesses. Companies like MakerBot and Formlabs have developed desktop **3D printers** that are affordable, easy to use, and capable of producing **high-quality objects**. This democratization of 3D printing has led to a surge in **innovation** across various fields, as people from all walks of life can now design and print their own creations.

Perhaps the most exciting development in **3D printing** is its potential in **healthcare**. 3D printing has been used to create custom **prosthetics, implants,** and even **bioprinted tissues.** In 2021, doctors in the Netherlands successfully implanted a 3D-printed **titanium skull** for a patient suffering from a life-threatening bone disorder. This level of precision and customization in medical devices would have been unthinkable just a decade ago. Furthermore, researchers are working on **3D printing organs** using a patient's own cells, which could one day eliminate the need for organ transplants and the complications that come with them.

Another area where 3D printing is making a significant impact is in **aerospace** and **automotive** manufacturing. Companies like SpaceX and Boeing are using 3D printing to produce **lightweight, complex components** that would be difficult or impossible to create using traditional methods. These parts not only reduce the weight of spacecraft and airplanes, but they are also faster and cheaper to produce. The ability to 3D print parts on demand is particularly valuable for space exploration, where transporting spare parts from Earth is costly and time-consuming.

Even the world of **fashion** is being transformed by 3D printing. Designers are using the technology to create intricate, **customizable** clothing and accessories that would be difficult to produce using traditional methods. Fashion brands are experimenting with 3D-printed shoes, jewelry, and even clothing, offering consumers a level of **personalization** never seen before. As the technology improves, it's likely that we'll see more 3D-printed fashion items making their way into mainstream retail.

Applications:
3D printing is no longer just for prototypes. It's now being used to create everyday objects and revolutionize various industries:

- **Healthcare:** 3D printing is changing the medical field by enabling the creation of custom **prosthetics, implants,** and even **bioprinted tissues.** Researchers are working on the ability to 3D print functional **organs,** which could eliminate the need for organ donors and reduce the risk of rejection in transplant patients. Customization at this level offers precision-tailored healthcare solutions.

- **Aerospace and Automotive:** Companies like SpaceX, Airbus, and Ford are using 3D printing to manufacture **complex parts** that reduce the weight and cost of spacecraft, airplanes, and cars. This innovation speeds up production timelines and allows for **on-demand parts**, which is especially useful in space missions where replacements must be manufactured on-site rather than transported from Earth.

- **Consumer Goods and Fashion:** The fashion industry is experimenting with **3D-printed designs**, from custom shoes and jewelry to full garments. 3D printing also allows consumers to create their own products at home using desktop printers, offering a new level of **personalization** and **sustainability**. As the technology improves, it will likely become an integral part of mainstream retail.

Summary:

Over the past decade, **3D printing** has evolved from a prototyping tool to a technology capable of producing complex, **functional objects** across various industries. From healthcare and aerospace to fashion and consumer goods, the applications of 3D printing are vast and growing. The ability to create **customized, on-demand products** with minimal waste is transforming how we think about manufacturing. As 3D printing technology continues to advance, we are likely to see even more innovative applications in areas we can only begin to imagine. With its **potential** to revolutionize everything from medicine to space exploration, 3D printing is shaping the future of design, manufacturing, and everyday life.

012 Exoplanet Discoveries
Searching for Habitable Worlds

For centuries, humans have gazed at the night sky and wondered if we are alone in the universe. The search for exoplanets—planets that orbit stars outside of our solar system—has brought us closer to answering that age-old question. Over the past decade, the discovery of thousands of **exoplanets** has reshaped our understanding of the cosmos and opened up the tantalizing possibility that some of these distant worlds could be **habitable**. The hunt for planets like Earth has never been more exciting, with advanced space telescopes and cutting-edge technology helping astronomers explore beyond our solar system in unprecedented ways.

The journey to finding exoplanets took a giant leap forward with the launch of **NASA's Kepler Space Telescope** in 2009. Kepler was designed to detect planets by observing the **transit method**—when a planet passes in front of its host star, causing a temporary dimming in the star's light. Over its nine-year mission, Kepler identified more than **2,600 exoplanets**, many of which were found in the **habitable zone**—the region around a star where conditions could be just right for liquid water to exist, a key ingredient for life as we know it. Among Kepler's discoveries were planets of all shapes and sizes, including **super-Earths** (planets larger than Earth but smaller than gas giants like Neptune) and **mini-Neptunes** (planets smaller than Neptune but larger than Earth).

One of the most significant milestones in the search for habitable worlds came in 2015 with the discovery of **Kepler-452b**, an exoplanet located in the habitable zone of a star similar to our Sun, about 1,400 light-years from Earth. Dubbed **"Earth's cousin"**, Kepler-452b is about 60% larger than Earth and may have conditions suitable for liquid water, raising the possibility that it could support life. While we still don't know if Kepler-452b has an **atmosphere** or **surface water**, its discovery highlighted the fact that there could be many Earth-like planets scattered throughout the galaxy, just waiting to be found.

Since Kepler's retirement in 2018, the search for exoplanets has been carried forward by newer missions like **NASA's Transiting Exoplanet Survey Satellite (TESS)**. Launched in 2018, TESS is designed to survey nearly the entire sky, focusing on **nearby stars**. TESS has already discovered **hundreds of exoplanets**, many of which are located close enough to Earth that future telescopes could analyze their atmospheres for signs of life. In fact, TESS has identified multiple **super-Earths** in the **habitable zones** of their stars, bringing us ever closer to finding a world that might harbor life.

The next leap in exoplanet discovery will come with the **James Webb Space Telescope (JWST)**, set to launch soon. This powerful space telescope is designed to study the atmospheres of exoplanets in unprecedented detail, using **infrared spectroscopy** to detect elements like water vapor, carbon dioxide, and methane—key indicators of habitability. JWST's ability to peer into the atmospheres of distant exoplanets could provide the first real clues about whether these worlds are capable of supporting life. The hope is that JWST will identify **biosignatures**, chemical markers in the atmosphere that could be produced by living organisms.

Applications:

The discovery of exoplanets has revolutionized our understanding of the universe and sparked new questions about the potential for life beyond Earth. Here's how these discoveries are advancing science and technology:

- **Searching for Life:** One of the main goals of exoplanet discovery is to find planets that could support life. With missions like **TESS** and the upcoming **James Webb Space Telescope**, scientists are now focusing on studying exoplanets' atmospheres to look for **biosignatures—gases** like oxygen, methane, or carbon dioxide that might indicate the presence of living organisms. Detecting these signs would be a monumental discovery in the search for extraterrestrial life.

- **Understanding Planetary Systems:** The study of exoplanets has taught us that planetary systems are incredibly diverse. Many of the exoplanets discovered are unlike anything in our solar system, ranging from **hot**

Jupiters (gas giants that orbit very close to their stars) to **super-Earths** and **mini-Neptunes**. This diversity has forced scientists to rethink how planetary systems form and evolve, and it offers valuable insights into the potential conditions for **life** in the universe.

- **Advancing Space Exploration Technology:** The hunt for exoplanets has driven the development of new technologies in space exploration. From **next-generation telescopes** like JWST to advances in **spectroscopy**, the search for **habitable worlds** is pushing the boundaries of what's possible in astronomy. These technologies not only help us discover new planets but also improve our understanding of the stars and galaxies beyond our own.

Summary:

The discovery of thousands of exoplanets over the past decade has transformed our view of the universe and brought us closer to answering one of humanity's oldest questions: Are we alone? With missions like **Kepler, TESS**, and the upcoming **James Webb Space Telescope**, we are finding more Earth-like planets in the habitable zones of distant stars, raising the possibility that some of these worlds could support life. The study of **exoplanets** is also reshaping our understanding of planetary systems and driving innovation in space exploration technology. As we continue to search the cosmos, the possibility of discovering a **habitable world**—and perhaps even **life**—seems more real than ever before.

013 Graphene
The Wonder Material of the Future

In 2004, scientists **Andre Geim** and **Konstantin Novoselov** at the University of Manchester isolated **graphene** for the first time, earning them the 2010 Nobel Prize in Physics. Since then, graphene has been hailed as the **"wonder material"** of the 21st century, with the potential to revolutionize industries ranging from electronics to medicine. Graphene is a single layer of carbon atoms arranged in a two-dimensional honeycomb lattice. What makes this material extraordinary is its combination of **properties**: it is the thinnest, strongest, lightest, and most conductive material known to science.

At just one atom thick, graphene is often described as a **2D material** because of its incredibly thin structure. Despite its **thinness**, graphene is remarkably **strong**—about 200 times stronger than steel, yet **flexible** and nearly transparent. It is also an **excellent conductor** of electricity and heat, making it ideal for applications in electronics, energy storage, composites, and biomedicine. The properties of graphene are so **versatile** that it has been called a **"miracle material"** with seemingly endless potential.

The isolation of graphene was achieved through an elegantly simple method: **mechanical exfoliation**, often referred to as the **"Scotch tape method"**. By peeling layers of graphite (the same material found in pencils) using adhesive tape, Geim and Novoselov were able to extract a single layer of carbon atoms—graphene. Since then, scientists have been developing more efficient ways to produce graphene at a larger scale, including **chemical vapor deposition (CVD)**, which involves growing graphene on metal surfaces using gases like methane.

One of the most exciting applications of graphene is in the field of **electronics**. Because graphene is so conductive and flexible, it has the potential to replace **silicon** in computer chips, transistors, and other electronic devices. Silicon has been the foundation of the electronics industry for decades, but it has physical limitations when it comes to miniaturization and speed. Graphene, on the other hand, could enable the development of **faster**, **smaller**, and **more energy-efficient** electronics. For example, researchers are exploring the use of graphene in flexible displays, wearable electronics, and even transparent touchscreens.

In addition to its use in electronics, graphene has also shown great promise in **energy storage**. Graphene-based **supercapacitors** could revolutionize the way we store and use energy by providing higher capacity and faster charging times compared to traditional batteries. This has significant implications for the development of **electric vehicles** and renewable energy technologies, where efficient energy storage is critical. Graphene **batteries**, for example, could dramatically increase the range of electric cars and reduce charging times, making electric vehicles more practical and accessible to the public.

Another area where graphene is making an impact is **biomedicine**. Due to its **biocompatibility** and strength, graphene is being used to develop **biosensors** for detecting diseases, as well as in the creation of **drug delivery systems** that can target specific cells in the body. Scientists are also exploring the use of graphene in **tissue engineering** to help regenerate damaged tissues, as its unique properties make it an excellent scaffold for cell growth.

Applications:

Graphene's versatility has made it one of the most exciting materials in modern science, with applications that span multiple industries:

- **Electronics and Computing:** Graphene's exceptional conductivity and flexibility make it a potential replacement for **silicon** in transistors, computer chips, and displays. Graphene-based electronics could be faster, smaller, and more energy-efficient, opening up new possibilities for flexible devices, transparent touchscreens, and even **quantum computing**.

- **Energy Storage:** In the field of energy, graphene is being used to create more **efficient batteries** and **supercapacitors**. Graphene-based batteries could significantly increase the **capacity of energy storage** devices, leading to longer-lasting electric vehicles and faster charging times. Additionally, graphene supercapacitors could enable quick bursts of energy for applications like **renewable energy** storage and **grid stabilization**.

- **Biomedicine:** The **biocompatibility** and flexibility of graphene make it a promising material for **medical applications**. Graphene is being used to develop **biosensors** for detecting diseases, as well as in **drug delivery systems** that can target specific cells in the body. Graphene-based materials are also being studied for use in **tissue engineering**, where they could help **regenerate** damaged tissues or support wound healing.

Summary:

Graphene is often called the **"wonder material"** of the future, and with good reason. Its combination of strength, flexibility, conductivity, and thinness makes it ideal for a wide range of applications, from **electronics** and **energy storage** to **biomedicine**. As research into graphene continues, the material's potential to revolutionize industries becomes clearer. Whether through creating faster, more efficient electronic devices, improving the storage of renewable energy, or advancing medical treatments, graphene has the power to reshape technology in the 21st century. The journey to fully harness its potential is still ongoing, but graphene is well on its way to transforming how we interact with the world around us.

014 Neuralink
Bridging the Brain-Technology Divide

Imagine a world where your brain can communicate directly with computers, where neurological conditions can be cured, and where human cognition is enhanced by technology. This futuristic vision is exactly what Neuralink, a company founded by Elon Musk in 2016, aims to achieve. Neuralink's mission is to develop **brain-computer interfaces (BCIs)** that would not only treat neurological disorders but also pave the way for **human-technology integration** at an unprecedented level. While this may sound like **science fiction**, recent advancements suggest that we are closer than ever to bridging the gap between the **human brain** and **machines**.

At its core, Neuralink is developing a technology that involves implanting tiny **electrodes** into the **brain**, which can both **read** and **stimulate** brain activity. These electrodes are connected to a chip, which then communicates wirelessly with external devices, such as computers or smartphones. The goal is to allow the brain to send and receive **information** directly from these devices, bypassing the need for physical inputs like keyboards or touchscreens. In the near term, Neuralink's technology could help patients suffering from conditions like **paralysis**, **Parkinson's disease**, or **epilepsy** by restoring lost functions through brain signals.

The technology behind **brain-computer interfaces** is based on decades of research into how the brain communicates through **electrical impulses**. Our neurons send and receive these impulses, forming the basis for all of our thoughts, movements, and sensations. BCIs work by detecting these signals and translating them into commands that **machines can understand**. However, the challenge has always been developing an interface that is both sensitive enough to read brain signals accurately and safe enough for **long-term implantation**. This is where Neuralink has made significant strides.

One of Neuralink's most impressive innovations is the development of **"threads"**—ultra-thin, flexible wires that are much thinner than a human hair. These threads can be implanted in the brain without causing significant damage to brain tissue. Each thread is equipped with multiple electrodes capable of both reading brain activity and stimulating the brain. These threads are connected to a small **implantable chip**, called the **Link**, which communicates wirelessly with external devices. The use of such fine, flexible materials is a major improvement over older BCI technologies, which relied on rigid electrodes that could cause scarring and limit the device's longevity.

In 2020, Neuralink demonstrated its technology in a live-streamed event, showcasing a pig named Gertrude who had a Neuralink implant in her brain. The device was able to record and display Gertrude's brain activity in real time as she moved around and interacted with her environment. While this was an early demonstration, it showed the potential of Neuralink's technology to track and interpret **complex brain activity**. The company's next goal is to test its technology in humans, starting with patients who are paralyzed, to help them control computers and smartphones with their minds.

Elon Musk has even more ambitious long-term goals for Neuralink. He envisions a future where humans can achieve **"symbiosis"** with **artificial intelligence (AI)**. Musk has frequently warned about the potential dangers of AI becoming more powerful than humans. Neuralink, he argues, could be the key to ensuring that humans stay ahead by merging with AI rather than being outpaced by it. While this idea is still far off, the development of BCIs could one day lead to enhanced cognitive abilities, memory, and even direct **brain-to-brain** communication, changing what it means to be human.

Applications:
While Neuralink's technology is still in the early stages, its potential applications are profound and far-reaching. Here are a few key areas where Neuralink could make a significant impact:

- **Restoring Motor Function:** Neuralink's most immediate application is in treating patients with **paralysis** or other **motor impairments**. By connecting the brain directly to computers or robotic limbs, Neuralink could enable paralyzed individuals to regain control over movement.

This could restore a significant degree of independence to people living with spinal cord injuries, ALS, or other neurological conditions.

- **Treating Neurological Disorders:** Another key area of focus for Neuralink is the treatment of **neurological disorders** like Parkinson's disease, epilepsy, and depression. By stimulating specific areas of the brain, Neuralink's implant could help regulate abnormal **brain activity**, reducing symptoms and improving quality of life. **Deep brain stimulation (DBS)**, a current treatment for Parkinson's, works on a similar principle, but Neuralink's device could offer more precision and fewer side effects.

- **Enhancing Human Cognition:** While this application is further off, Musk has stated that one of Neuralink's long-term goals is to **enhance human cognition** by integrating our brains with AI. This could involve improving memory, learning speed, and even problem-solving **capabilities**. Such enhancements could allow humans to process information faster and more efficiently, making us more competitive in a world where AI is becoming increasingly powerful.

Summary:
Neuralink is pioneering a new frontier in technology with its development of **brain-computer interfaces (BCIs)**, aiming to connect the human brain directly with machines. By implanting ultra-thin **electrodes** in the brain, Neuralink's device can both read and stimulate brain activity, potentially restoring lost functions in patients with **neurological disorders** like paralysis and Parkinson's disease. In the long term, Neuralink envisions a future where humans can achieve symbiosis with AI, enhancing cognitive abilities and even bridging the gap between biological and artificial intelligence. While these ambitions are still in their early stages, the potential of Neuralink to revolutionize **medicine** and **human-technology** interaction is undeniable. As the technology develops, it could reshape how we think, move, and interact with the world around us.

015 Wearable Tech
Health Monitoring on Your Wrist

In the past decade, **wearable technology** has gone from a novelty to an integral part of everyday life, with millions of people relying on devices like **smartwatches** and **fitness trackers** to monitor their health and wellness. What was once seen as a trend for tech enthusiasts has now become a powerful tool in personal healthcare, with the ability to track everything from heart rate and sleep patterns to detecting irregularities that could indicate serious health conditions. The rise of **wearable tech** has revolutionized how we manage our health, placing cutting-edge technology quite literally at our fingertips—or more accurately, on our **wrists**.

The most popular form of wearable technology is the **smartwatch**, pioneered by companies like Apple, Fitbit, and Garmin. These devices go beyond simply telling time; they are equipped with **sensors** that can monitor various aspects of a **wearer's health**, such as heart rate, blood oxygen levels, and physical activity. For many, wearing a device like the Apple Watch or a Fitbit has become a daily routine, offering real-time insights into their **health** and motivating them to stay active.

One of the most significant advancements in wearable tech has been its ability to **track cardiovascular health**. Early smartwatches primarily tracked steps and basic heart rate information, but newer models are capable of monitoring more detailed metrics, such as **heart rate variability (HRV)**, which can indicate stress levels, and **electrocardiograms (ECGs)**, which can detect irregular heartbeats known as **arrhythmias**. The Apple Watch, for instance, is equipped with an ECG app that can alert users to potential issues like **atrial fibrillation (AFib)**, a condition that increases the risk of stroke. By providing users with this kind of information, wearable tech can empower individuals to take proactive steps in managing their health or seek medical attention before a condition worsens.

Wearable tech's potential isn't just limited to heart health. Devices like the Fitbit Sense and the Oura Ring are pushing the boundaries of what health **data** can be **collected** and **analyzed**. These devices can monitor **sleep patterns**, track respiratory rate, and even measure **blood oxygen saturation (SpO2)**, which has become particularly important during the COVID-19 pandemic. By analyzing this data, wearables can help users optimize their **sleep**, manage **stress**, and track overall wellness. The ability to gather continuous data on health metrics allows for a more complete picture of a person's well-being over time, enabling early detection of issues that may not be immediately apparent.

But wearable tech is not just about passive monitoring—it's also about **active intervention**. Many wearables now offer personalized health insights and recommendations based on the data they collect. For example, smartwatches can provide reminders to move if the user has been sitting for too long, or offer guided breathing exercises when elevated stress levels are detected. In some cases, these devices can even integrate with **healthcare systems**, allowing users to share their health data with their doctors for more informed **diagnoses** and **treatments**.

Wearable technology is also evolving to help manage **chronic conditions**. Devices like the Freestyle Libre and Dexcom **continuous glucose monitors (CGMs)** have transformed the way people with **diabetes** manage their condition. These wearable sensors continuously monitor blood **sugar levels** and send **real-time data** to the user's smartphone, eliminating the need for finger-prick tests. This data helps individuals with diabetes manage their insulin levels more effectively, reducing the risk of complications and improving their quality of life.

Looking ahead, the future of wearable tech in healthcare is incredibly promising. Companies are already developing **wearable blood pressure monitors, non-invasive glucose sensors**, and even wearable **thermometers** that could provide continuous monitoring of core body temperature. As artificial intelligence (AI) becomes more integrated with wearables, the data collected could be analyzed to provide even more personalized health insights, detecting patterns that users may not notice themselves.

Applications:

Wearable tech is transforming personal healthcare by providing real-time monitoring and insights into various aspects of health and wellness. Here are some of the key applications:

- **Cardiovascular Health:** Smartwatches equipped with **heart rate monitors** and **ECG** capabilities can detect irregular heart rhythms like **atrial fibrillation (AFib)** and provide early warnings of potential cardiovascular issues. This allows users to seek medical intervention before serious problems arise, reducing the risk of conditions like **stroke**.

- **Chronic Condition Management:** Devices like **continuous glucose monitors (CGMs)** have revolutionized the way people with **diabetes** manage their blood sugar levels. By providing continuous, real-time data, these wearables enable better **insulin** management and help prevent dangerous fluctuations in blood sugar.

- **Wellness and Sleep Monitoring:** Wearables like the Fitbit Sense and Oura Ring track **sleep patterns, respiratory rate**, and **blood oxygen levels**, helping users optimize their overall wellness. By identifying sleep disturbances or stress, wearables can guide users toward healthier habits and more restorative sleep.

Summary:

Wearable technology has emerged as a **powerful tool** in personal healthcare, allowing users to monitor their **health** in **real time** and take proactive steps to improve their well-being. From tracking heart health with **ECG**-capable smartwatches to managing chronic conditions like **diabetes** with continuous glucose monitors, wearables are providing personalized insights that can detect **health issues** early and help prevent serious complications. As wearable tech continues to evolve, with the potential to monitor even more health metrics, it's clear that these devices are playing an increasingly important role in how we manage our **health**. Whether you're tracking your sleep, managing stress, or monitoring your heart, wearable tech is revolutionizing health monitoring—right on your **wrist**.

016 Solar Energy Breakthroughs
Powering the Future

As the world grapples with the challenges of climate change and the need to reduce our reliance on fossil fuels, **solar energy** has emerged as one of the most promising solutions to our energy needs. Over the past decade, breakthroughs in **solar technology** have made it **more efficient**, affordable, and accessible than ever before. Once seen as a niche energy source, solar power is now transforming into a mainstream method of generating electricity, and it's playing a crucial role in the transition to a **clean, renewable energy future**.

Solar energy works by converting sunlight into electricity through the use of **photovoltaic (PV) cells**, which are typically made from **silicon**. When sunlight hits these cells, it excites **electrons**, creating an **electric current** that can be captured and used to power everything from homes and businesses to entire cities. What makes solar energy so attractive is its **abundance**—the Earth receives more energy from the sun in just one hour than the entire world uses in a year. Harnessing even a fraction of this energy could meet the world's energy needs many times over.

In the past, the high cost and relatively low efficiency of **solar panels** were major obstacles to the widespread adoption of solar energy. However, in the last decade, there have been significant **advancements** in **solar technology** that have improved both the efficiency of solar panels and their affordability. The cost of solar power has dropped by more than 80% since 2010, thanks in part to improvements in manufacturing processes, economies of scale, and advances in **materials science**. Today, solar power is not only cost-competitive with fossil fuels, but in many parts of the world, it is now cheaper than coal and natural gas.

One of the most exciting breakthroughs in solar energy is the development of **perovskite solar cells**. Unlike traditional silicon-based solar cells, **perovskites** are a class of materials that can be manufactured more cheaply and easily, while offering **higher efficiency**. Perovskite solar cells have achieved efficiency rates of over 25% in laboratory settings, and they hold the potential to surpass silicon in commercial applications. These cells can also be applied to flexible surfaces, opening up possibilities for solar panels that can be integrated into windows, walls, and even clothing.

Another major advancement in solar technology is the development of **bifacial solar panels**, which can capture sunlight from both sides of the panel. Traditional solar panels only capture sunlight that hits their front surface, but bifacial panels can also **capture light** that is reflected off the ground or surrounding surfaces, increasing their energy output by up to 30%. This innovation makes solar installations more efficient, especially in environments with reflective surfaces, such as deserts or snowy regions.

Energy storage has long been one of the biggest challenges for solar power, since the sun doesn't shine 24/7. However, breakthroughs in **battery technology** are helping to solve this problem. The development of more efficient and affordable **lithium-ion batteries**, along with next-generation energy storage solutions, is enabling solar power to be stored and used even when the sun isn't shining. Companies like Tesla are at the forefront of this effort, with products like the Powerwall, which allows homeowners to store excess solar energy generated during the day for use at night. These advancements in storage are making solar energy a more reliable and consistent power source, reducing the need for backup from fossil fuels.

Another innovation in solar energy is the rise of **solar farms** and **community solar programs**, which allow large-scale solar energy production and provide access to solar power for people who might not be able to install panels on their own homes. **Utility-scale solar farms** are now producing enough electricity to power entire cities, while community solar projects allow residents to subscribe to solar energy from a shared array, making renewable energy accessible to renters, apartment dwellers, and low-income households.

Applications:
The advancements in solar energy have wide-reaching applications that are transforming the way we generate and use electricity:

- **Residential Solar Power:** With the cost of **solar panels** continuing to decline, more homeowners are installing solar panels on their rooftops, reducing their dependence on the traditional power grid and lowering their electricity bills. **Innovations** like the Tesla Powerwall are allowing homeowners to store excess energy, making solar a **viable solution** even in areas with inconsistent sunlight.

- **Utility-Scale Solar Farms:** Large-scale **solar farms** are becoming a significant source of electricity in many regions. These farms use vast arrays of solar panels to generate electricity for thousands of homes and businesses, reducing reliance on fossil fuels. The growth of **bifacial panels** and **perovskite cells** is making solar farms more efficient and cost-effective, enabling greater adoption of clean energy.

- **Energy Storage Solutions:** The development of **advanced batteries** and **energy storage systems** is solving one of solar energy's biggest challenges: **intermittency**. By storing solar power generated during the day for use at night or during cloudy periods, energy storage technologies are making solar a more reliable source of power, reducing the need for fossil fuel backup and supporting the integration of renewables into the power grid.

Summary:
Recent breakthroughs in **solar energy** have revolutionized the industry, making it more affordable, efficient, and accessible than ever before. From the development of **perovskite solar cells** and **bifacial panels** to advancements in **energy storage technology**, solar power is becoming a key player in the transition to a **clean energy future**. Whether through residential rooftop installations, utility-scale solar farms, or innovative energy storage solutions, solar energy is providing a **sustainable alternative** to fossil fuels, reducing greenhouse gas emissions, and helping to combat climate change. As the cost of solar continues to decline and technology improves, solar energy is poised to play a central role in powering the future.

017 AI in Art
Machines That Create Masterpieces

For centuries, art has been seen as one of the most human forms of expression—a creative process that requires imagination, emotion, and a deep understanding of the world around us. But in recent years, the boundaries between human creativity and machine intelligence have begun to blur, thanks to the rise of **artificial intelligence (AI)** in the world of art. **AI-generated art** has become a powerful and controversial force in the creative world, with machines now capable of producing **paintings, music, poetry,** and even **films**. This blending of **technology and creativity** is pushing the limits of what we consider art and challenging traditional notions of authorship and originality.

At the **heart of AI** in art is **machine learning**, particularly a branch known as **deep learning**. Using vast amounts of data, **AI algorithms** can be trained to recognize patterns and generate new content that mimics human-created works. One of the most popular approaches is the use of **Generative Adversarial Networks (GANs)**. GANs consist of two neural networks: one generates new images, while the other evaluates them. The **two networks work together**, improving the quality of the generated art until the result is nearly indistinguishable from human-created pieces. Through this process, **AI systems** have been able to generate incredibly detailed and unique works of art, sometimes even surpassing what a human artist might create in terms of complexity and precision.

One of the most famous examples of AI-generated art is the 2018 sale of a painting titled **"Portrait of Edmond de Belamy"**, created by an **AI algorithm** developed by the Paris-based art collective Obvious. The portrait, which depicts a fictional aristocrat in an abstract, dreamlike style, was created using a GAN trained on thousands of portraits from art history. The painting sold at auction for a staggering $432,500, sparking a debate about whether **AI-generated art** can truly be considered "art" and who, if anyone, should be credited as the artist—the algorithm, its creators, or both?

AI is also making waves in **music composition**. AI algorithms like OpenAI's **MuseNet** and **AIVA** have been trained on massive datasets of classical music and can now compose original pieces in the style of great composers like **Bach, Mozart,** and **Beethoven**. These **AI-composed pieces** are so convincing that even seasoned musicians sometimes struggle to tell them apart from human-composed music. In addition, AI is being used in more modern genres of music, with algorithms capable of generating everything from jazz to electronic dance music, pushing the boundaries of creativity in the music industry.

The **rise of AI** in **art** also extends to **poetry** and **literature**. OpenAI's GPT-3, a powerful language model, can generate everything from poems and short stories to articles and essays that are often indistinguishable from those written by humans. These AI-generated texts can be deeply creative, thought-provoking, and stylistically unique, blurring the line between human creativity and machine intelligence. Artists and writers are now collaborating with AI systems to **co-create** literary works, opening up new possibilities for creative expression.

But AI's influence on art goes beyond just creating new works. AI is also being used to analyze, preserve, and restore existing art. For example, **AI algorithms** can be trained to analyze the brushstrokes, color palettes, and techniques of famous painters, helping art historians and **conservationists** understand how certain works were created and even restore damaged or incomplete pieces to their former glory. This use of AI in the art world is providing new insights into the history of art and allowing us to preserve our cultural heritage for future generations.

Despite the excitement surrounding **AI-generated art**, its rise has also sparked significant debates and controversies. Critics argue that **artificial intelligence** lacks the emotional depth, intentionality, and consciousness that are fundamental to human creativity. Others question whether AI art can truly be original, given that **AI models** are trained on existing works of art and rely on patterns found in those works to create new pieces. These debates raise important questions about **authorship**, **creativity**, and the role of technology in artistic expression.

Applications:
The impact of AI in art is being felt across multiple creative fields, from visual arts to music and literature. Here are some key applications:

- **Visual Arts:** AI-generated art, created using technologies like **Generative Adversarial Networks (GANs)**, is producing paintings, sculptures, and other visual works that are often indistinguishable from human-made art. These works are being displayed in galleries and even sold at auctions, challenging traditional concepts of artistic creation and authorship.

- **Music Composition:** AI is also making significant contributions to music composition. Algorithms like AIVA and MuseNet can generate original music in the style of classical composers or modern genres, offering composers new tools for collaboration and creativity. AI is being used to compose soundtracks, symphonies, and even pop songs.

- **Creative Writing and Poetry:** AI systems like GPT-3 are capable of generating original poems, short stories, and articles that mimic human writing. These **AI tools** are being used by authors and poets to co-create literary works, allowing for new forms of **collaborative creativity** that blend human input with machine-generated text.

Summary:
AI in art is revolutionizing the creative process, enabling machines to generate everything from visual masterpieces to music and poetry. Through technologies like deep learning and **Generative Adversarial Networks (GANs)**, AI systems can analyze vast amounts of artistic data and produce original works that challenge our traditional understanding of creativity. While **AI-generated art** has sparked debates about **authorship, originality**, and the nature of art, it has also opened up exciting new possibilities for **collaboration between humans and machines**. As AI continues to evolve, its role in the world of art will only grow, creating masterpieces that blend technology and creativity in ways we are only beginning to understand.

018 Blockchain and Cryptocurrencies
Revolutionizing Finance

In just over a decade, **blockchain technology** and **cryptocurrencies** have gone from obscure concepts to global forces that are transforming the world of finance. Originally created as the underlying technology for **Bitcoin**, blockchain is now being used to **revolutionize** everything from banking and payments to supply chain management and even voting systems. Meanwhile, cryptocurrencies have opened the door to a new **decentralized financial system** that challenges traditional institutions and gives individuals more control over their money. Together, blockchain and cryptocurrencies are reshaping how we think about trust, transactions, and the future of finance.

At its core, **blockchain** is a **distributed ledger technology** that allows data to be recorded and shared across a network of computers in a secure, transparent, and immutable way. Unlike traditional databases, which are controlled by a **central authority**, a blockchain is decentralized—meaning no single entity has control over the data. Instead, transactions are recorded in blocks and linked together in a chain, with each block verified by a **network of computers** (called **nodes**) through a process known as **consensus**. This makes blockchain incredibly secure, as altering any information in a blockchain would require changing every subsequent block in the chain, which is virtually impossible.

The most famous application of blockchain technology is **Bitcoin**, the world's first cryptocurrency, which was introduced in 2009 by an anonymous entity known as **Satoshi Nakamoto**. Bitcoin uses blockchain to create a

decentralized digital currency that allows users to send and receive payments without the need for a trusted intermediary like a **bank**. Transactions on the Bitcoin blockchain are verified by **miners**, who use powerful computers to solve complex mathematical problems, securing the network in exchange for newly minted Bitcoins.

Since the creation of Bitcoin, thousands of other cryptocurrencies have been developed, each with its own unique features and use cases. **Ethereum**, for example, introduced the concept of **smart contracts**, which are self-executing contracts where the terms are written directly into code. This allows for automated, trustless transactions that can be used for everything from financial agreements to **decentralized applications (dApps)**. Other cryptocurrencies, such as **Ripple** and **Litecoin**, offer faster transaction times or more energy-efficient consensus mechanisms, expanding the possibilities of what blockchain technology can do.

One of the most significant promises of **blockchain** and **cryptocurrencies** is their ability to create a decentralized financial system—often referred to as **DeFi (decentralized finance)**. DeFi platforms, built primarily on Ethereum's blockchain, enable users to access financial services like lending, borrowing, and trading without relying on traditional financial institutions like **banks**. By removing intermediaries, DeFi offers faster, cheaper, and more accessible financial services to anyone with an internet connection, regardless of their location. This **democratization of finance** has the potential to bring banking services to the billions of people around the world who are currently **unbanked** or **underbanked**.

Blockchain's ability to provide transparency and security has also made it attractive to industries beyond finance. In **supply chain management**, for example, blockchain is being used to track goods from production to delivery, ensuring **authenticity** and **reducing fraud**. Companies like Walmart and IBM have implemented blockchain to trace food products, improving transparency and safety in the **supply chain**. Meanwhile, **smart contracts** are being used to automate complex legal agreements, insurance claims, and even real estate transactions, reducing the need for intermediaries and cutting down on time and costs.

Despite its potential, blockchain and cryptocurrencies are not without challenges. The **volatility** of cryptocurrencies like Bitcoin has made them less attractive as a stable store of value, while the energy-intensive process of mining has raised environmental concerns. Governments and regulators are also grappling with how to manage this new form of money, with debates over how to balance innovation with the need to protect consumers and prevent illegal activities like **money laundering** and **fraud**. **Scalability** is another issue, as many blockchain networks struggle to handle large numbers of transactions quickly and efficiently, limiting their ability to compete with traditional payment systems like Visa or Mastercard.

However, the **potential benefits** of blockchain and cryptocurrencies continue to drive innovation and investment in the space. Major companies like Tesla, PayPal, and Square have begun accepting Bitcoin and other cryptocurrencies, while countries like **El Salvador** have even adopted Bitcoin as legal tender. Central banks are also exploring the development of their own digital currencies (known as **CBDCs**, or **central bank digital currencies**), which would combine the benefits of blockchain technology with the stability of government-backed currency. These developments suggest that blockchain and cryptocurrencies are here to stay, and their role in the future of finance will only grow.

Applications:
The rise of blockchain technology and cryptocurrencies is revolutionizing multiple industries, particularly finance. Here are some key applications:

- **Decentralized Finance (DeFi):** DeFi platforms enable users to access financial services such as lending, borrowing, and trading without the need for **traditional intermediaries** like banks. Built on blockchain networks like Ethereum, DeFi allows for faster, cheaper, and more inclusive financial transactions, giving users **full control** over their assets.

- **Cross-Border Payments:** Cryptocurrencies like Bitcoin and Ripple offer a faster, **cheaper alternative** to traditional cross-border payment systems, which are often slow and costly due to intermediaries and foreign exchange fees. Blockchain technology streamlines these transactions, making it easier for individuals and businesses to send money internationally.

- **Supply Chain Management:** Blockchain is transforming supply chain management by providing **transparency** and **traceability**. Companies can use blockchain to track goods at every stage of the supply chain, ensuring **authenticity** and reducing **fraud**. This has applications in industries such as food safety, pharmaceuticals, and luxury goods.

Summary:
Blockchain technology and **cryptocurrencies** are revolutionizing the world of finance by creating a decentralized, secure, and transparent system for transactions. From the rise of **Bitcoin** and **Ethereum** to the growth of **DeFi** platforms, blockchain is reshaping the way we think about money, trust, and financial services. The technology's ability to remove intermediaries, reduce costs, and provide greater transparency has far-reaching implications.

019 Augmented Reality
Merging Real and Digital Worlds

Augmented reality (**AR**) has long been a staple of science fiction, but in recent years, it has become a reality that is transforming the way we interact with both the physical and digital worlds. Unlike **virtual reality** (**VR**), which immerses users in a fully digital environment, AR overlays **digital content**—images, information, or interactive elements—onto the real world in real-time. This fusion of the digital and physical realms is revolutionizing industries from gaming and entertainment to education, healthcare, and retail, opening up endless possibilities for how we experience the world around us.

The rise of **smartphones** and **wearable devices** has brought AR into the mainstream. Millions of people were first introduced to augmented reality with the **global sensation** Pokémon GO, a mobile game released in 2016 that allowed players to **catch virtual creatures** overlaid onto real-world environments using their phone's camera. This game demonstrated the potential of AR to **merge** the real and digital worlds in a fun and engaging way, and it quickly became a cultural phenomenon. Since then, AR has evolved far beyond gaming, finding practical applications in a wide range of industries.

One of the most exciting developments in AR is its ability to enhance the way we **learn and work**. In the field of **education**, AR is being used to create **immersive learning experiences** that bring subjects to life in ways that were previously impossible. For example, students studying **biology** can use AR apps to explore the human body in 3D, viewing organs, bones, and muscles as if they were right in front of them. Similarly, engineering students can use AR to visualize **complex machinery** and **designs**, allowing them to understand intricate systems in a hands-on, interactive way. This type of experiential learning makes abstract concepts more tangible, increasing engagement and improving **comprehension**.

In the workplace, AR is transforming industries like **manufacturing**, **logistics**, and **healthcare** by providing workers with real-time, context-specific information. **AR glasses** like Microsoft HoloLens and Google Glass allow workers to see digital overlays on physical objects, guiding them through tasks such as assembly, repairs, or maintenance. For example, an assembly line worker could use AR to see step-by-step instructions overlaid onto the parts they are working with, **reducing errors** and **speeding up** the process. In healthcare, AR is being used in **surgical procedures**, allowing surgeons to visualize patient anatomy in 3D while performing complex operations, improving precision and reducing risks.

Retail is another industry that has embraced AR as a way to enhance the shopping experience. Retailers like IKEA and Sephora are using AR to let customers **visualize products** before making a purchase. With the IKEA Place app, for example, customers can use their smartphone camera to see how a piece of furniture would look in their home before they buy it, helping them make more informed decisions. Similarly, beauty brands like Sephora use AR to allow customers to try on makeup **virtually**, using their phone's camera to see how different shades of lipstick or eyeshadow would look on their face. These types of AR experiences are making shopping more **interactive**, **personalized**, and **convenient**.

The entertainment industry continues to be a key driver of **AR innovation**. Beyond gaming, AR is being used in live events, sports, and even television to create more **immersive** experiences for audiences. Sports broadcasts, for example, often use AR to overlay statistics, player information, or replays onto the field, enhancing the viewer's experience. Similarly, AR is being used in concerts and live performances to add digital elements that interact with the physical stage, creating visually stunning shows that blend the real and digital worlds.

Despite its rapid adoption and impressive advancements, AR is still in its **early stages**, and there are many challenges that need to be addressed for it to reach its full potential. Issues like **hardware limitations**, **battery life**, and the need for faster, more reliable **internet connections** (such as 5G) are all hurdles that AR developers are working to overcome. Additionally, creating seamless, user-friendly AR experiences requires significant improvements in **software development** and user interface design. However, as technology continues to evolve, the potential for AR to become an integral part of our daily lives grows stronger.

Applications:
Augmented reality is making its mark in a wide range of industries, providing practical solutions and enhancing user experiences in new and innovative ways:

- **Education and Training:** AR is transforming education by creating **immersive learning experiences**. Students can explore 3D models, engage with interactive content, and gain a deeper understanding of complex subjects, from anatomy to engineering. In the workplace, AR is being used to train employees with real-time instructions overlaid onto physical tasks, improving efficiency and reducing mistakes.

- **Retail and E-Commerce:** AR is enhancing the shopping experience by allowing customers to **visualize products** in their own environment before making a purchase. From furniture to makeup, AR apps are making it easier for customers to make informed decisions and enjoy personalized shopping experiences.

- **Healthcare and Surgery:** In the medical field, AR is being used to provide surgeons with real-time, **3D visualizations** of patient anatomy during procedures. This technology enhances **precision** and reduces **risks**, particularly in complex surgeries. AR is also being used for medical training, giving students the ability to practice on virtual patients before performing procedures on real ones.

Summary:
Augmented reality (AR) is merging the real and digital worlds in ways that are transforming industries and enhancing everyday life. From the success of Pokémon GO to practical applications in **education**, **healthcare**, **retail**, and more, AR is providing **immersive**, **interactive** experiences that bridge the gap between physical and digital realities. As the technology continues to evolve, with improvements in hardware, software, and connectivity, the

potential for AR to become a seamless part of our daily lives is becoming increasingly likely. Whether you're learning, shopping, or working, AR is opening up new possibilities for how we interact with the world around us.

020 Space Tourism
The Next Frontier for Travel

For decades, the idea of **space tourism** seemed like something out of science fiction, a far-off dream reserved for astronauts and sci-fi movies. But in the 21st century, that dream is rapidly becoming a reality. Thanks to **advancements** in **technology** and the ambitious visions of companies like SpaceX, Blue Origin, and Virgin Galactic, the concept of **commercial space travel** is now within reach for civilians. Space tourism is set to become the **next frontier for travel**, offering a select few the opportunity to venture beyond Earth's atmosphere and experience the wonders of space firsthand.

The modern era of space tourism began in 2001, when American businessman Dennis Tito became the first private citizen to travel to space. Tito paid a reported $20 million to fly aboard a Russian Soyuz spacecraft and spend several days on the **International Space Station (ISS)**. This landmark event proved that private individuals could, in fact, venture into space, sparking interest in **space tourism** as a **viable industry**. However, for nearly two decades, space tourism remained the domain of a few ultra-wealthy adventurers who could afford the staggering **costs**.

Fast forward to today, and the dream of space tourism is closer than ever to becoming a commercial reality. Companies like Virgin Galactic and Blue Origin are developing **suborbital spaceflights** designed to give civilians the chance to experience a few minutes of weightlessness and see the Earth from the edge of space. In July 2021, Richard Branson, founder of Virgin Galactic, and Jeff Bezos, founder of Blue Origin, each successfully flew to space aboard their companies' respective spacecrafts—VSS Unity and New

Shepard. These historic flights were not just about the experience of the individuals on board; they marked the beginning of a **new era** in which **commercial space travel** is accessible to more than just astronauts.

One of the key distinctions of **suborbital flights**, like those offered by Virgin Galactic and Blue Origin, is that they allow passengers to reach space without completing a full orbit around Earth. These flights typically last between 10 and 15 minutes, during which passengers experience **zero gravity** for a brief period and are treated to breathtaking views of **Earth's curvature** from over 50 miles above the surface. While short in duration, these experiences are truly transformative, offering a **once-in-a-lifetime** perspective that only a handful of people have ever experienced.

While **suborbital flights** are the first step, companies like SpaceX have their sights set on even more ambitious **space tourism** ventures—orbital flights and beyond. SpaceX, founded by Elon Musk, has already demonstrated its capability to send astronauts to the ISS with its Crew Dragon spacecraft. Now, the company is developing plans to take private citizens on multi-day orbital missions around Earth and even to the Moon. In 2021, SpaceX made history with the launch of Inspiration4, the first **all-civilian space mission**, in which four private individuals orbited Earth for three days aboard the Crew Dragon. This mission not only showcased the possibilities of civilian space travel but also raised millions of dollars for charity, highlighting the potential for **space tourism** to have a broader impact on society.

The ultimate goal of companies like SpaceX is to enable **interplanetary travel**. Elon Musk has been vocal about his vision of establishing a human settlement on **Mars**, and he sees space tourism as a critical step toward making humanity a **multiplanetary species**. While trips to **Mars** may still be a distant reality, the groundwork is being laid today for **longer-duration space missions** that could take private citizens to the **Moon**, **Mars**, and beyond.

While the prospect of space tourism is thrilling, it also comes with significant **challenges** and **questions**. One of the most immediate concerns is the **cost**. Current **ticket prices** for suborbital flights range from $200,000 to $500,000, making space tourism an experience reserved for the **wealthy**. However, as technology advances and competition increases, it's expected that the cost of space travel will decrease over time, potentially making it more accessible to a broader audience. There are also **environmental concerns**, as the rocket launches required for space tourism contribute to carbon emissions. Companies like SpaceX and Blue Origin are investing in **reusable rockets** to make space travel more sustainable, but the environmental impact remains a **key issue** that needs to be addressed as the industry grows.

Moreover, space tourism raises important questions about **regulation** and **safety**. While companies like SpaceX, Virgin Galactic, and Blue Origin have made significant strides in ensuring the safety of their spacecraft, space travel is **inherently risky**. Ensuring that the proper regulations and oversight are in place to **protect passengers** and the environment will be crucial as the industry continues to evolve.

Applications:
While space tourism is still in its early stages, it has the potential to revolutionize the travel industry and inspire new technological advancements. Here are some key applications:

- **Suborbital Space Travel:** Companies like Virgin Galactic and Blue Origin are offering **suborbital flights** that allow passengers to experience a brief period of **weightlessness** and see the **Earth** from the edge of space. These short flights are making space accessible to private citizens for the first time, providing a unique travel experience.

- **Orbital Space Missions:** SpaceX is leading the charge in orbital space tourism, offering **multi-day missions** that take civilians around Earth. Future plans include sending private citizens on missions to the **Moon** and even **Mars**, pushing the boundaries of space exploration and travel.

- **Inspiration and Innovation:** Space tourism has the potential to inspire a **new generation of explorers and innovators**. By making space travel more accessible, it could fuel interest in **STEM fields (science, technology, engineering, and math)** and lead to new technologies that benefit humanity as a whole.

Summary:
Space tourism is no longer a distant **dream**—it is becoming a **reality**, thanks to the efforts of companies like SpaceX, Virgin Galactic, and Blue Origin. From **suborbital flights** that offer a few minutes of **weightlessness** to ambitious plans for orbital missions around **Earth** and trips to the **Moon** and **Mars**, space tourism is set to revolutionize travel and exploration. While challenges like **cost, safety,** and **environmental impact** remain, the

potential for space tourism to inspire new technologies and make space more accessible to civilians is immense. As the next frontier for travel, space tourism promises to change the way we see and experience the **universe**.

021 Advanced Prosthetics
Merging Biology with Technology

In recent years, **advanced prosthetics** have made incredible strides, transforming the lives of millions of people who have lost limbs or suffered debilitating injuries. With the help of **cutting-edge technologies**, modern prosthetics are doing more than just replacing lost limbs—they are merging **biology** with **technology**, enabling users to regain mobility, dexterity, and even sensory feedback. These advancements are not only helping individuals return to their daily lives but are also pushing the boundaries of what's possible in human augmentation.

Traditional prosthetics have been used for centuries to help amputees regain some functionality. However, older models were often simple, rigid, and limited in their movement. The development of **advanced prosthetics** has changed that dramatically. Today's prosthetic limbs are equipped with **robotic components**, **biomechanical sensors**, and **AI-driven algorithms** that allow them to mimic the natural movement of human limbs. These prosthetics are often custom-fitted to each user, providing a more comfortable and functional experience than ever before.

One of the most groundbreaking innovations in modern prosthetics is the development of **myoelectric prosthetics**. These prosthetic limbs use **electrical signals** generated by the user's muscles to control the movement of the limb. Sensors embedded in the prosthetic socket detect these signals, allowing users to move their prosthetic limb simply by thinking about the movement, just as they would with a biological limb. This technology gives amputees greater control over their movements, making tasks like picking up objects, typing, or even shaking hands much more intuitive and natural.

Another remarkable advancement in prosthetics is the integration of **bionic technology**. **Bionic limbs** are designed to work with the body's nervous system, allowing the user to control the prosthetic through direct neural connections. In some cases, sensors implanted in the muscles or nerves can detect electrical impulses from the brain, enabling the prosthetic to move in response to those signals. For example, T**argeted Muscle Reinnervation (TMR)** is a surgical technique that reassigns nerves from an amputated limb to other muscles, allowing the prosthetic to interpret neural signals with greater precision. This technology not only improves movement control but also enhances the user's sense of ownership over their prosthetic limb.

Beyond mobility, one of the most exciting developments in prosthetics is the ability to provide **sensory feedback**. For many amputees, the loss of sensation is one of the most challenging aspects of using a prosthetic limb. However, researchers are now developing prosthetics that can "feel" through the use of **haptic feedback systems**. These systems use sensors and actuators to mimic the sensation of touch, sending signals back to the user's nervous system. This allows the user to feel **pressure**, **texture**, and **temperature** through their prosthetic, greatly improving the functionality and user experience of the limb. In 2020, researchers successfully demonstrated a **bionic arm** that could restore a sense of touch, giving users the ability to feel objects in real-time.

The impact of these advancements is profound. For individuals who have lost limbs, **advanced prosthetics** offer more than just functionality—they provide a sense of **independence** and **normalcy**. Veterans, athletes, and everyday individuals are now able to return to activities they once thought were impossible. Athletes like Paralympians are using advanced prosthetics to compete at the highest levels, while others are benefiting from custom-designed prosthetics that allow them to walk, run, and even rock climb with confidence and precision.

One of the most exciting possibilities for the future of advanced prosthetics is the merging of **artificial intelligence (AI)** with prosthetic design. **AI-driven prosthetics** are being developed to learn from the user's movements, allowing the prosthetic to anticipate and adapt to their needs in real-time. For example, AI can analyze the way a user walks and adjust the prosthetic's response to improve **balance**, **speed**, and **efficiency**. These **smart prosthetics** are not only reactive but predictive, enhancing the user's overall mobility and quality of life.

As prosthetics continue to evolve, the line between biology and technology is becoming increasingly blurred. In the future, prosthetics may be so advanced that they surpass the capabilities of biological limbs. **Biohybrid limbs**, which combine **biological tissue** with **robotic components**, are already in development, offering the potential for even more seamless integration between the body and technology. This could lead to a new era of human augmentation, where prosthetics are not only used to replace lost limbs but to enhance human abilities beyond what is naturally possible.

Applications:

The advancements in prosthetics are having a profound impact on the lives of amputees and individuals with physical disabilities. Here are some key applications:

- **Myoelectric Prosthetics:** Using **electrical signals** from the user's muscles, **myoelectric prosthetics** allow for intuitive control of the prosthetic limb. This technology has enabled users to perform everyday tasks like gripping, lifting, and typing with greater ease and precision.
- **Bionic Limbs:** Bionic technology connects prosthetics directly to the body's nervous system, allowing for more precise movement control. Techniques like **Targeted Muscle Reinnervation (TMR)** enhance the user's ability to control their prosthetic, creating a seamless interaction between the body and the artificial limb.

- **Sensory Feedback:** Advanced prosthetics are now being developed with **haptic feedback systems** that allow users to feel pressure, texture, and temperature through their prosthetic limbs. This sensory feedback greatly improves the functionality of prosthetics and gives users a more natural experience.

Summary:

Advanced prosthetics are merging **biology with technology**, providing amputees and individuals with physical disabilities the opportunity to regain control, mobility, and even sensory feedback. From **myoelectric prosthetics** that allow for intuitive movement to **bionic limbs** that connect directly to the nervous system, the advancements in prosthetic technology are transforming lives. The integration of **artificial intelligence (AI)** and **haptic feedback** is pushing the boundaries of what prosthetics can achieve, offering users not only functional limbs but enhanced capabilities. As this technology continues to evolve, the future of prosthetics promises to be one where human ability is augmented and redefined.

022 Plastic-Eating Enzymes
Tackling the Global Waste Problem

The world is drowning in plastic. Every year, **over 300 million tons** of plastic are produced globally, with a significant portion ending up in landfills, oceans, and natural ecosystems. This **non-biodegradable material** can take hundreds of years to break down, causing long-term environmental damage. As plastic waste continues to accumulate, the search for **innovative solutions** has become more urgent than ever. Enter plastic-eating enzymes—a **revolutionary discovery** that could help combat the global plastic waste problem by breaking down plastic at a molecular level, offering a promising solution for one of the world's most pressing environmental challenges.

The breakthrough in plastic-eating enzymes came in **2016** when scientists discovered a **bacterium** called **Ideonella sakaiensis** in a Japanese recycling plant. This bacterium was found to **feed on polyethylene terephthalate (PET)**, a common type of plastic used in everything from water bottles to food packaging. The bacterium produces a pair of **enzymes, PETase and MHETase**, that work together to break down PET plastic into its basic building blocks, which can then be reused to create new plastic products. This discovery was the first of its kind, showing that **nature** might already have a solution to the plastic waste crisis.

Following this initial discovery, researchers began working to enhance the **efficiency** of these **plastic-eating enzymes**. In 2018, scientists at the **University of Portsmouth** in the UK engineered a mutant version of PETase that was able to degrade plastic even faster than the natural enzyme. This breakthrough opened the door to further research into how enzymes could be used on a large scale to break down plastics and recycle them in an environmentally friendly way. The goal is to **develop enzymes** that can be used in **industrial-scale recycling plants**, where they can break down vast quantities of plastic waste efficiently and economically.

What makes **plastic-eating enzymes** so promising is their ability to break down plastic into its original components, which can then be reused to create new plastic products. This is in stark contrast to traditional mechanical recycling, where plastics are melted down and reformed into lower-quality products. With enzyme-based recycling, the plastic can be fully broken down into its **monomers**, allowing for closed-loop recycling—where plastic waste is turned back into **high-quality, reusable** plastic. This process reduces the need for virgin plastic production, which relies on fossil fuels, and could significantly decrease the amount of plastic waste that ends up in the environment.

In addition to **PET**, researchers have been exploring enzymes that can break down **other types of plastic**. For example, scientists have discovered enzymes that can degrade **polyurethane (PU)**, a type of plastic used in insulation, furniture, and footwear. Polyurethane is particularly challenging to recycle, as it releases toxic chemicals when broken down using traditional methods. However, the development of enzymes capable of breaking down PU in a safe and efficient way could provide a **sustainable solution** for managing this difficult-to-recycle plastic.

One of the most exciting advancements in this field is the development of **enzyme cocktails—combinations** of different enzymes that can break down a wider range of plastics more efficiently. By combining enzymes like **PETase** and **MHETase** with other plastic-degrading enzymes, researchers hope to create a powerful tool for tackling the diversity of plastic waste in the environment. These enzyme cocktails could be used in waste management facilities to process mixed plastic waste, breaking it down into **reusable components** without the need for complex sorting and separation processes.

While the potential of **plastic-eating enzymes** is immense, there are still challenges to overcome before they can be deployed on a large scale. One major hurdle is the **cost** of producing these enzymes in large quantities. Currently, enzyme production is expensive, and scaling up to the level needed to tackle the global plastic waste problem will require significant advances in **biotechnology** and **industrial production** methods. Additionally, researchers are working to improve the **stability** and **efficiency** of these enzymes in **real-world conditions**, as the environment in which plastic waste accumulates (landfills, oceans, etc.) can be harsh and unpredictable.

Despite these challenges, the discovery of **plastic-eating enzymes** represents a major step forward in the fight against plastic pollution. Governments, corporations, and environmental organizations are all taking notice of this technology's potential. In 2020, Carbios, a French biotech company, successfully demonstrated the use of an enzyme-based recycling process in an industrial setting, breaking down **PET** waste into its raw components and using them to create new plastic bottles. This successful trial marked an important milestone in bringing enzyme-based plastic recycling closer to commercial reality.

Applications:

The development of plastic-eating enzymes has the potential to revolutionize how we manage and recycle plastic waste. Here are some of the key applications:

- **Enzyme-Based Recycling:** PETase and other plastic-eating enzymes could be used in recycling facilities to break down plastic waste into its monomers, allowing for **closed-loop recycling**. This process would reduce the need for virgin plastic production and minimize plastic pollution by turning waste into reusable materials.

- **Degrading Hard-to-Recycle Plastics:** Enzymes that can break down plastics like **polyurethane (PU)** offer a sustainable solution for dealing with plastics that are difficult to recycle using traditional methods. This could help address the problem of toxic chemicals released during traditional recycling processes and create more **eco-friendly methods** of disposal.

- **Environmental Cleanup:** In the future, **enzyme cocktails** could be deployed to break down plastic waste in the environment, such as **ocean plastic**. By speeding up the natural degradation of plastic, these enzymes could help reduce the long-term environmental damage caused by plastic pollution in oceans, rivers, and other ecosystems.

Summary:

The discovery of **plastic-eating enzymes** offers a promising solution to the global plastic waste crisis. By breaking down plastics like **PET** and **polyurethane** at a molecular level, these enzymes enable **closed-loop recycling**, turning plastic waste back into high-quality, reusable materials. While challenges like **cost** and **scaling** remain, advancements in enzyme efficiency and the development of **enzyme cocktails** are bringing us closer to deploying this technology on an industrial scale. As the world seeks sustainable ways to manage plastic waste, **plastic-eating enzymes** represent a groundbreaking step toward a cleaner, greener future.

023 AI in Healthcare
Diagnosing Diseases with Precision

Artificial intelligence (AI) is transforming industries across the globe, but perhaps nowhere is its impact more profound than in the field of **healthcare**. From **early diagnosis** to **treatment planning** and **predictive analytics**, AI is revolutionizing how diseases are detected, managed, and treated. With its ability to analyze vast amounts of data and identify patterns that might elude even the most experienced clinicians, AI is helping healthcare professionals diagnose diseases with a level of **precision** and **speed** that was previously unimaginable. This breakthrough in healthcare technology is poised to save countless lives and improve the quality of care for patients worldwide.

At the heart of AI's role in healthcare is its ability to process and **analyze medical data** at an unprecedented scale. AI algorithms, particularly those based on **machine learning** and **deep learning**, can be trained on vast datasets of patient records, medical images, and genetic information. These algorithms learn to recognize patterns that indicate the presence of diseases, often with a level of accuracy that rivals or surpasses human doctors. For example, AI systems are being used to analyze **medical imaging** such as **X-rays, CT scans,** and **MRIs**, identifying early signs of conditions like **cancer, cardiovascular disease**, and **neurological disorders** with remarkable precision.

One of the most significant breakthroughs in AI healthcare applications has been in the detection of **cancer**. AI-powered tools are now capable of identifying tumors in **mammograms, lung scans,** and other medical images at a stage so early that they might be missed by the human eye. In 2020,

Google Health's **AI system** demonstrated the ability to detect breast cancer in mammograms more accurately than radiologists, reducing both false positives and false negatives.
This kind of precision allows for earlier intervention, which is critical in improving patient outcomes and survival rates.

Another area where AI is making a profound impact is in the diagnosis and management of **cardiovascular diseases**. AI algorithms can analyze data from wearable devices, like **smartwatches** and **fitness trackers**, to detect abnormal heart rhythms or changes in heart rate variability that could indicate conditions like **atrial fibrillation (AFib)**. Early detection of AFib is crucial in preventing strokes, and AI is helping doctors identify these irregularities before they lead to more serious complications. Additionally, AI systems are being used to assess **heart scans**, predict the likelihood of **heart attacks**, and even recommend personalized treatment plans based on a patient's unique risk factors.

AI is also transforming the field of **genomics**. By analyzing genetic data, AI can help identify mutations and variations in a person's DNA that might be linked to specific diseases. This has opened up new possibilities in the diagnosis and treatment of **rare genetic disorders** that are often difficult to detect through traditional methods. AI-powered genomic analysis can also play a crucial role in **personalized medicine**, where treatment plans are tailored to a patient's genetic makeup, ensuring more effective and targeted therapies.

One of the most exciting developments in AI healthcare is its ability to analyze **electronic health records (EHRs)** and predict a patient's likelihood of developing certain diseases. By processing large datasets of patient records, AI can identify patterns and risk factors that might not be immediately apparent to doctors. This predictive capability allows healthcare providers to intervene early and develop preventive strategies for patients at high risk of developing conditions like **diabetes**, **chronic kidney disease**, or **hypertension**. AI's ability to **forecast disease** progression can also help doctors adjust treatment plans in real-time, improving patient outcomes.

In addition to diagnosing diseases, AI is playing a growing role in **treatment planning** and **drug discovery**. By analyzing medical literature, patient data, and clinical trials, AI systems can recommend personalized treatment options based on the latest research and the patient's unique profile. AI is also accelerating the development of new drugs by identifying promising **compounds** and predicting how they might interact with the human body. This has the potential to dramatically reduce the time and cost involved in bringing new treatments to market.

However, despite the incredible promise of AI in healthcare, there are still **challenges** to overcome. **Data privacy** and **security** are major concerns, as the use of AI requires access to sensitive patient information. Ensuring that AI systems are trained on diverse datasets is also crucial, as **biased data** can lead to inaccurate diagnoses or recommendations, particularly for underrepresented populations. Additionally, the integration of AI into **healthcare systems** requires a balance between automation and human oversight, ensuring that AI augments rather than replaces the expertise of healthcare professionals.

Despite these challenges, the future of AI in healthcare is **incredibly promising**. Major tech companies like Google, IBM, and Microsoft, alongside startups and research institutions, are investing heavily in **AI healthcare solutions**. As these technologies continue to evolve and improve, we are likely to see even more sophisticated tools that enhance diagnostic accuracy, speed up treatment, and ultimately save lives.

Applications:

AI in healthcare is having a transformative impact across various fields of medicine. Here are some key applications:

- **Medical Imaging and Early Diagnosis:** AI-powered tools are being used to analyze **X-rays, CT scans**, and **MRIs** to detect early signs of diseases like **cancer** and **cardiovascular conditions**. These tools can identify abnormalities with high precision, allowing for earlier and more accurate diagnosis.

- **Wearable Health Monitoring:** AI algorithms are being integrated into **wearable devices** like **smartwatches**, which can monitor heart rate, blood pressure, and other vital signs. These devices can detect irregularities in real-time, helping doctors intervene before conditions like **AFib** lead to serious complications.

- **Genomic Analysis:** AI is being used to analyze **genetic data**, identifying mutations and markers linked to **genetic disorders**. This technology is critical in the field of **personalized medicine**, where treatments are tailored to a patient's genetic profile, offering more effective therapies.

Summary:

Artificial intelligence (AI) is revolutionizing healthcare by enabling the **early detection** and **precise diagnosis** of diseases that might otherwise go unnoticed. From analyzing **medical images** to detecting **genetic mutations** and monitoring **heart health** through wearable devices, AI is helping doctors diagnose diseases with unparalleled accuracy, a new era of precision medicine that will save lives and improve health outcomes for patients around the world.

024 Quantum Cryptography
Securing the Future of Data

In an age where **cybersecurity threats** are growing in both sophistication and frequency, the need for more secure ways to protect sensitive information has never been more critical. Enter **quantum cryptography**—a revolutionary approach to data security that harnesses the principles of **quantum mechanics** to create unbreakable encryption. While traditional cryptographic methods are vulnerable to being cracked by increasingly powerful computers, quantum cryptography promises to secure the future of data, providing a level of protection that is virtually impossible to hack. As quantum computers become more advanced, **quantum cryptography** is emerging as a vital solution to keep our communications, financial transactions, and sensitive information safe.

At the heart of **quantum cryptography** is a concept known as **quantum key distribution (QKD)**. Unlike classical encryption, where data is secured by complex mathematical algorithms that could potentially be broken by brute force, **QKD** leverages the strange and powerful properties of **quantum particles** to secure data. Specifically, **QKD** uses **quantum bits**, or **qubits**, to transmit encryption keys between two parties. These qubits are transmitted as **photons**—particles of light—over a fiber optic cable or through free space. The beauty of **quantum cryptography** lies in the **quantum properties of qubits**, which make them incredibly secure.

One of the key principles of **quantum mechanics** is the concept of **superposition**, where a quantum particle like a photon can exist in multiple states simultaneously. Another key principle is **entanglement**, where two or more particles become correlated in such a way that the state of one particle instantly affects the state of the other, no matter how far apart they are. These principles create a level of complexity that is virtually impossible to replicate or intercept. Even more important for cryptography is the **observer effect**, which states that any attempt to measure or eavesdrop on a quantum system will disturb the system and alter the data being transmitted. This means that if a hacker tries to intercept the encryption key, the act of eavesdropping would change the key and alert the sender and receiver to the intrusion.

This **unique property** of **quantum mechanics** is what makes **QKD** so powerful. With **quantum key distribution**, the encryption keys used to encode and decode messages are transmitted in a way that guarantees their security. Any attempt to intercept or tamper with the key would be immediately detected, ensuring that only the intended recipient can access the information. This level of security is unachievable with classical cryptographic systems, where powerful computers or algorithms can theoretically crack encryption codes given enough time and processing power.

Quantum cryptography is already moving out of the realm of theory and into practical applications. In 2017, China launched the world's first **quantum satellite**, known as **Micius**, which successfully demonstrated the feasibility of **quantum key distribution** over long distances. This experiment showed that **QKD** could be used to securely transmit encryption keys between two locations more than 1,200 kilometers apart, a breakthrough in the field of secure communications. Since then, countries around the world have been racing to develop their own **quantum networks**, with governments, banks, and defense agencies at the forefront of these efforts.

One of the most significant applications of **quantum cryptography** is in the area of **financial transactions**. With the rise of online banking, e-commerce, and digital payments, securing financial data has become a **top priority**. Quantum cryptography offers a way to ensure that sensitive financial information, such as credit card numbers, account details, and personal identification, can be transmitted safely and securely, without the risk of being intercepted by **cybercriminals**. As traditional encryption methods become increasingly vulnerable to hacking, quantum cryptography could become the standard for protecting financial transactions and keeping customers' data safe.

Another **critical application** of quantum cryptography is in securing **government communications** and **military data**. Governments around the world are investing heavily in quantum cryptography to protect classified information and ensure the security of their communications networks. The ability to transmit encryption keys securely is especially important in the context of national security, where the interception of sensitive information could have dire consequences. Quantum cryptography provides an **unbreakable defense** against **cyber espionage**, offering a level of protection that is critical in today's geopolitical landscape.

However, while **quantum cryptography** holds immense promise, there are still challenges that need to be overcome before it can be widely adopted. One of the primary challenges is **scalability**. Currently, **QKD systems** require **expensive** and **specialized** equipment, including **quantum repeaters** that extend the range of **quantum communications**. Scaling quantum networks to cover large geographical areas, like entire cities or countries, is a significant technical hurdle that researchers are working to solve. Additionally, while quantum cryptography is theoretically unbreakable, implementing it in real-world systems requires overcoming issues like signal degradation over long distances and maintaining the **integrity** of **qubits** during transmission.

Despite these challenges, the future of **quantum cryptography** looks incredibly promising. As **quantum computers** continue to develop, the need for more advanced cryptographic systems will become even more critical. Classical encryption, such as **RSA** and **AES**, which currently protect most online communications, could be rendered obsolete by quantum computers capable of solving complex problems that are impossible for today's computers. This potential threat has spurred a global effort to develop **post-quantum cryptography**, with quantum cryptography emerging as the most secure solution for the **post-quantum era**.

Applications:
The development of quantum cryptography offers revolutionary potential for securing sensitive data across various industries. Here are some of its most critical applications:

- **Financial Transactions:** As financial institutions increasingly rely on digital transactions, quantum cryptography could ensure that **credit card numbers, banking details**, and **customer data** are transmitted with an unprecedented level of security. This could prevent cyberattacks on financial systems and protect users' personal information.

- **Government and Military Communications:** Governments and defense agencies are adopting **quantum cryptography** to secure classified communications and sensitive information. By using **quantum key distribution,** they can protect critical data from cyber espionage and ensure the integrity of national security systems.

- **Healthcare Data Security:** With the rise of **electronic health records (EHRs)** and telemedicine, quantum cryptography could play a crucial role in securing patient data and ensuring that **medical information** remains confidential and tamper-proof during transmission.

Summary:

Quantum cryptography is revolutionizing the field of data security by using the principles of **quantum mechanics** to create encryption systems that are virtually unbreakable. Through **quantum key distribution (QKD),** data can be transmitted securely, with any attempt to intercept the encryption keys being instantly detected. From **financial transactions** and **government communications** to **healthcare data,** quantum cryptography offers a level of protection that is unmatched by classical cryptographic methods. As the development of **quantum computers** continues to advance, **quantum cryptography** will play a crucial role in securing the future of data in the post-quantum world.

025 Hydrogen Fuel Cells
Clean Energy for Transportation

As the world faces the urgent challenge of reducing greenhouse gas emissions and transitioning to **clean energy, hydrogen fuel cells** are emerging as a promising solution for the future of **transportation**. With the potential to provide **zero- emission** power for vehicles, hydrogen fuel cells are seen as a key technology for decarbonizing the transportation sector. Unlike traditional combustion engines that burn fossil fuels and release harmful emissions, **hydrogen fuel cells** generate electricity through a clean chemical reaction, emitting only water vapor as a byproduct. As the race to find sustainable alternatives to gasoline and diesel intensifies, hydrogen fuel cells are poised to play a critical role in the transition to a **clean energy future**.

At the heart of **hydrogen fuel cells** is a simple yet powerful chemical process: **electrochemical conversion**. A hydrogen fuel cell generates electricity by **combining hydrogen** and **oxygen** in a **chemical reaction** that produces electricity, heat, and water. The process begins when **hydrogen gas (H_2)** is fed into the **anode** of the fuel cell. At the anode, a catalyst causes the hydrogen molecules to **split** into **protons** and **electrons**. The protons pass through a membrane to the cathode, while the electrons are forced to take an external circuit, generating an electric current. When the protons and electrons meet with **oxygen (O_2)** at the cathode, they form **water (H_2O)** as a harmless byproduct. This clean, efficient process makes hydrogen fuel cells a **zero-emission energy source**, perfect for powering vehicles without contributing to air pollution or climate change.

One of the most exciting applications of **hydrogen fuel cells** is in **transportation**, where they can provide a clean alternative to internal combustion engines. **Fuel cell electric vehicles (FCEVs)** operate similarly to **battery electric vehicles (EVs)**, but instead of storing energy in a battery, FCEVs generate electricity on demand using hydrogen. This gives them several key advantages over traditional electric vehicles, particularly in terms of **range** and **refueling time**. While battery electric vehicles need to be recharged over several hours, hydrogen fuel cell vehicles can be refueled in just a few minutes, similar to gasoline-powered cars. Additionally, FCEVs typically have a longer driving range than battery electric vehicles, making them an attractive option for long-distance travel.

One of the leading companies in the development of hydrogen fuel cell vehicles is Toyota, with its **Toyota Mirai** being one of the first commercially available FCEVs. The Mirai can travel up to **400 miles** on a single tank of hydrogen and **emits** nothing but **water vapor**, making it one of the **cleanest cars** on the road today. Other major automakers, including Hyundai and Honda, are also investing in hydrogen-powered vehicles, with models like the Hyundai Nexo and Honda Clarity joining the growing lineup of fuel cell cars. These vehicles offer the convenience of fast refueling and long range, making them well-suited for both personal and commercial use.

But the potential of **hydrogen fuel cells** extends far beyond passenger cars. Hydrogen is also being explored as a clean energy solution for heavier modes of **transportation**, such as **buses, trucks, trains**, and even **ships**. For example, companies like Nikola and Hyundai are developing hydrogen-powered trucks that can travel long distances without emitting pollutants, offering a sustainable alternative for the freight and logistics industry. Similarly, countries like Germany are already using hydrogen trains to replace diesel-powered locomotives, reducing emissions on regional rail lines. The maritime industry is also looking at hydrogen as a way to power ships, helping to decarbonize the global shipping sector.

One of the **biggest advantages** of hydrogen fuel cells is their ability to provide clean, **renewable energy** without relying on fossil fuels. Hydrogen can be produced from a variety of sources, including renewable energy like wind and solar power, through a process known as **electrolysis**. In this process, electricity from renewable sources is used to split water into **hydrogen** and **oxygen**, producing green hydrogen that can be used in fuel cells. This makes hydrogen an ideal way to store excess energy from renewable sources and use it to power vehicles, homes, or entire cities when needed. By integrating hydrogen fuel cells into a broader clean energy ecosystem, we can create a **sustainable energy infrastructure** that reduces our dependence on fossil fuels and cuts carbon emissions.

However, there are still challenges to overcome before hydrogen fuel cells can be widely adopted. One of the main hurdles is the **cost of hydrogen production**, particularly green hydrogen, which is still more expensive than hydrogen produced from natural gas. Another challenge is the need for a comprehensive **hydrogen refueling infrastructure**. While there are growing networks of hydrogen stations in regions like California, Europe, and Japan, much of the world lacks the infrastructure needed to support widespread hydrogen vehicle adoption. Governments, automakers, and energy companies are investing heavily in expanding hydrogen refueling networks, but it will take time to build out the necessary infrastructure.

Despite these challenges, the **future** of hydrogen fuel cells looks **bright**. Governments around the world are recognizing the potential of hydrogen to play a key role in the transition to clean energy. Countries like Germany, Japan, and South Korea are leading the way with national hydrogen strategies that aim to expand the use of hydrogen across multiple sectors, including transportation, industry, and power generation. In the United States, the Biden administration has also highlighted hydrogen as a key component of its climate strategy, with plans to invest in hydrogen infrastructure and research.

Applications:
Hydrogen fuel cells are revolutionizing transportation and offering a clean, efficient solution for powering vehicles. Here are some of the key applications:

- **Passenger Vehicles: Fuel cell electric vehicles (FCEVs)** like the **Toyota Mirai** and Hyundai Nexo provide clean, zero-emission alternatives to gasoline-powered cars, with fast refueling times and long driving ranges. These vehicles are well-suited for everyday driving and long-distance travel.

- **Heavy-Duty Transportation:** Hydrogen fuel cells are being used to power **trucks, buses, trains**, and **ships**, offering a sustainable solution for industries that rely on heavy-duty vehicles. Hydrogen-powered trucks and buses are being developed to reduce emissions in freight and public transportation.

- **Energy Storage:** Hydrogen can be produced from **renewable energy** sources and used as a clean energy carrier, storing excess wind or solar power and converting it into electricity through fuel cells when needed. This integration of hydrogen into renewable energy systems supports a sustainable **energy infrastructure**.

Summary:

Hydrogen fuel cells are emerging as a key technology in the transition to **clean energy**, offering zero-emission power for vehicles and other applications. Through the process of electrochemical conversion, hydrogen fuel cells generate electricity, heat, and water, providing a clean alternative to fossil fuels. With **fuel cell electric vehicles (FCEVs)** like the Toyota Mirai and Hyundai Nexo, as well as applications in heavy-duty transportation and energy storage, hydrogen fuel cells are playing a critical role in decarbonizing the transportation sector. While challenges like the cost of green hydrogen and infrastructure development remain, hydrogen fuel cells have the potential to **revolutionize transportation** and pave the way for a **clean energy** future.

026 AI-Driven Drug Discovery
Finding New Cures Faster

Artificial intelligence (AI) is transforming industries across the board, but one of its most profound impacts is being felt in the field of **drug discovery**. Traditional drug development is a complex, time-consuming, and expensive process, often taking **10 to 15 years** and billions of dollars to bring a new drug to market. However, **AI-driven drug discovery** is revolutionizing this process, allowing researchers to find potential cures and treatments at an unprecedented pace. By leveraging AI's ability to analyze massive amounts of data, identify patterns, and predict outcomes, pharmaceutical companies are now able to **accelerate the discovery** of new drugs, reduce costs, and bring life-saving treatments to patients faster than ever before.

The traditional process of drug discovery involves identifying potential **biological targets**, screening thousands of compounds, and testing them in the lab. This can take years, as researchers often have to sift through vast amounts of data and conduct extensive **trial-and-error experiments** to find a promising candidate. AI is changing this paradigm by using **machine learning algorithms** and **deep learning** to automate and optimize much of the process. These AI systems can quickly analyze enormous datasets, including **genomic information**, **chemical properties**, and **clinical trial data**, to identify potential drug candidates with far greater speed and accuracy than humans alone.

One of the most significant breakthroughs in **AI-driven drug discovery** came during the **COVID-19 pandemic**. In the race to find treatments and vaccines, researchers turned to AI to identify **promising drug candidates**. For example, companies like BenevolentAI and Insilico Medicine used AI to screen existing drugs for their potential to treat COVID-19. These **AI systems**

analyzed thousands of molecules, looking for compounds that could inhibit the virus's ability to replicate. As a result, drugs like Baricitinib, which was originally developed to treat rheumatoid arthritis, were identified as potential treatments for COVID-19 in **record time**. This rapid identification and repurposing of existing drugs showcased AI's ability to streamline drug discovery in the face of urgent **global health crises**.

Beyond identifying existing drugs for new uses, AI is also helping scientists discover **entirely new compounds**. Traditionally, researchers would rely on their knowledge of **chemical structures** and **biological pathways** to design new molecules, often working with limited information. AI, however, can analyze millions of chemical compounds and predict how they will interact with biological targets. This has led to the discovery of novel molecules that might never have been considered through traditional methods. In 2020, the first **AI-discovered drug, DSP-1181**, entered clinical trials. Developed by the UK-based company Exscientia, this drug, designed to treat **obsessive-compulsive disorder (OCD)**, was created using AI to optimize the chemical structure of the molecule. The entire process, from initial discovery to clinical trials, took less than 12 months, compared to the typical 4 to 6 years in traditional drug discovery.

One of the key strengths of **AI-driven drug discovery** is its ability to **analyze big data**. Modern drug discovery generates vast amounts of data from multiple sources, including **genomic sequencing**, **high-throughput screening**, and **clinical trials**. AI can sift through this data, identifying **biomarkers**, **genetic mutations**, and other factors that are associated with disease. By understanding these underlying mechanisms, AI can help researchers design more targeted therapies. For example, AI is being used to identify **precision medicines** for cancer patients by analyzing genetic data to determine which drugs are most likely to be effective for specific individuals. This approach not only improves the chances of successful treatment but also reduces the risk of **side effects**.

Another exciting development in AI-driven drug discovery is the use of **predictive modeling** to forecast how drugs will behave in the human body. By simulating the **pharmacokinetics** and **pharmacodynamics** of a drug, AI can predict how it will be absorbed, distributed, metabolized, and excreted. This helps researchers identify potential safety issues early in the process, reducing the risk of failure in later stages of development. AI can also predict potential **drug-drug interactions**, ensuring that new therapies can be safely combined with existing treatments. This capability is particularly valuable in fields like oncology, where patients often receive multiple treatments simultaneously.

The integration of **AI in drug discovery** is not limited to the pharmaceutical industry. Research institutions and universities are also adopting **AI tools** to **accelerate** their work. In 2020, the AI system AlphaFold, developed by DeepMind, made headlines for solving one of biology's greatest challenges: **predicting protein folding**. Proteins are the building blocks of life, and understanding how they fold into specific 3D shapes is critical for drug design. AlphaFold's ability to **predict protein structures** with near-experimental accuracy is revolutionizing drug discovery, providing researchers with crucial insights into how drugs interact with their targets at the molecular level.

While AI-driven **drug discovery** holds enormous promise, there are still challenges to overcome. One of the biggest hurdles is the need for **high-quality data**. AI systems rely on large datasets to make accurate predictions, and any biases or gaps in the data can lead to suboptimal results. Additionally, while AI can accelerate the early stages of drug discovery, **clinical trials** and **regulatory** approval processes remain time-consuming and expensive. Ensuring that AI-driven discoveries can make it through these later stages will require continued collaboration between AI developers, pharmaceutical companies, and regulatory agencies.

Despite these challenges, the future of AI in drug discovery looks incredibly promising. Major pharmaceutical companies like Pfizer, Novartis, and Roche are investing heavily in AI technologies, while startups and tech giants like Google and IBM Watson are leading the charge in developing **AI-driven drug discovery platforms**. As AI continues to evolve and improve, it is poised to reshape the pharmaceutical industry, bringing **life-saving treatments** to patients faster and more efficiently than ever before.

Applications:
AI-driven drug discovery is revolutionizing the pharmaceutical industry, offering faster, more efficient ways to find new treatments and cures. Here are some key applications:

- **Drug Repurposing**: AI can analyze existing drugs to identify **new uses** for them. During the COVID-19 pandemic, AI systems were used to find drugs like **Baricitinib** that could be repurposed to treat the virus, significantly speeding up the development of treatments.

- **Novel Drug Discovery:** AI is being used to design new molecules and discover novel compounds. The first AI-discovered drug, **DSP-1181**, entered clinical trials in 2020, demonstrating AI's potential to rapidly accelerate the discovery process.

- **Precision Medicine:** AI is helping researchers develop **precision therapies** by analyzing genetic data to determine which drugs are most effective for specific patients. This personalized approach improves treatment outcomes and reduces side effects.

Summary:
AI-driven drug discovery is revolutionizing the way new treatments are discovered, developed, and brought to market. By harnessing the power of **machine learning** and **big data analysis**, AI is accelerating the drug discovery process, identifying promising compounds, and optimizing drug design. From **repurposing existing drugs** to finding **novel compounds**, AI is enabling researchers to develop treatments faster and with greater **precision**. As AI continues to evolve, its role in precision medicine, clinical trials, and pharmacokinetics will only grow, promising a future where life-saving treatments can be developed in **record time** and tailored to individual patients. The integration of AI into drug discovery is not just speeding up the process—it's transforming the future of medicine.

027 Smart Homes
Living with Artificial Intelligence

Artificial intelligence (AI) has made its way into nearly every aspect of our lives, but one of the most personal and transformative places it has taken root is in our homes. **Smart homes**, powered by AI, are redefining the way we live, turning ordinary houses into **intelligent environments** that can anticipate our needs, automate tasks, and improve our quality of life. From smart thermostats and voice assistants to AI-driven security systems and energy management, smart homes are at the forefront of the **AI revolution**, making daily life more convenient, efficient, and secure.

The concept of a **smart home** revolves around the integration of connected devices—often referred to as the **Internet of Things (IoT)**—that communicate with each other to create a seamless, automated living experience. AI takes this a step further by **analyzing data** from these devices, learning from user behavior, and making decisions in real time. For example, AI-enabled systems can learn your daily routine and **automatically** adjust the thermostat, lighting, or blinds to optimize comfort and energy efficiency, all without you lifting a finger. This shift from manual control to intelligent automation is transforming homes into **smart ecosystems** that are more responsive to their inhabitants' needs.

One of the most visible and widely adopted AI technologies in smart homes is the **voice assistant**. Devices like Amazon Alexa, Google Assistant, and Apple's Siri have become the central hubs of many **smart homes**, allowing users to control their environment with simple **voice commands**. Whether it's adjusting the temperature, turning on the lights, playing music, or ordering groceries, these **AI-driven assistants** offer a hands-free way to manage a wide range of tasks. Over time, these systems learn users' preferences, becoming more **intuitive** and able to anticipate needs, making them invaluable in daily life.

Smart homes aren't just about convenience—they're also about **safety** and **security**. AI-driven security systems are helping homeowners **monitor** and **protect** their **property** with more precision than ever before. **Smart security cameras**, like those from Ring or Nest, use AI to detect motion, recognize faces, and differentiate between potential threats and harmless activity, such as a dog walking by. These systems can send **real-time alerts** to homeowners, allowing them to monitor their property remotely from their smartphones. Additionally, AI-powered **doorbell cameras** and **smart locks** provide secure access to the home, enabling homeowners to see who's at the door and grant or deny entry from anywhere in the world. This integration of AI into home security gives homeowners **peace of mind**, knowing their property is being monitored by intelligent systems that can react quickly and accurately to potential threats.

Another critical application of AI in smart homes is **energy management**. As concerns about climate change and energy consumption grow, smart homes are playing a key role in reducing energy waste. **AI-powered thermostats**, like Google's Nest or Ecobee, learn the homeowner's schedule and adjust the temperature to minimize energy usage without sacrificing comfort. These systems can even optimize energy use by taking into account weather forecasts and the thermal characteristics of the house. In addition, smart appliances like **washing machines**, **refrigerators**, and **dishwashers** can be programmed to run during off-peak energy hours, further reducing energy costs and environmental impact. This ability to intelligently **manage energy consumption** makes smart homes more sustainable, lowering utility bills and reducing the household's carbon footprint.

Beyond the essentials of comfort, security, and energy management, AI is also bringing **health** and **wellness** into the home. **Smart mirrors** and **fitness trackers** can monitor health metrics like heart rate, sleep patterns, and activity levels, providing personalized insights into a resident's well-being. Some AI-driven systems can even suggest workouts, recommend dietary changes, or track medication schedules to help homeowners maintain a healthy lifestyle. For example, **smart beds** equipped with sensors can monitor sleep quality and adjust mattress firmness or temperature to improve sleep conditions. As these technologies continue to advance, the integration of **AI into home health** could play a vital role in preventive healthcare, enabling individuals to track and manage their health more effectively from the comfort of their own home.

Entertainment is another area where AI is enhancing the smart home experience. AI-powered systems can learn users' preferences for **music**, **movies**, and **TV shows**, curating personalized playlists and recommendations across streaming services like Netflix, Spotify, or YouTube. In some cases, AI can even analyze the user's mood based on their interactions and suggest content accordingly. **Smart speakers** and **TVs** are becoming integral parts of the home entertainment system, offering immersive audio and visual experiences that are tailored to each family member's tastes.

As AI in **smart homes** continues to evolve, the future promises even more advanced and seamless integration of technologies. The concept of **ambient intelligence**—where AI systems are embedded into the environment and operate **autonomously** without user input—is rapidly becoming a reality. Imagine a home that not only adjusts to your needs automatically but also **predicts your future needs** based on patterns of behavior and data from multiple devices. AI-driven homes could one day become fully **self-regulating environments** that manage everything from energy and water usage to meal planning and home maintenance, all while optimizing for efficiency, sustainability, and comfort.

However, the rise of **AI-powered smart homes** does come with challenges, particularly in the areas of **privacy** and **security**. As more data is collected from smart home devices, ensuring the privacy of homeowners and protecting against **cyberattacks** becomes a critical concern. Smart homes rely on a constant exchange of data between devices, often through the cloud, making them potential targets for hackers. Companies developing smart home technology are working to strengthen **data encryption**, improve **user authentication**, and build more robust firewalls to protect against these threats. As smart homes become more connected, finding a balance between convenience and security will be essential.

Applications:

AI-powered smart homes are revolutionizing the way we live by integrating intelligent systems into every aspect of daily life. Here are some key applications:

- **Home Automation and Convenience:** AI systems like Amazon Alexa and Google Assistant allow homeowners to **control everything** from lights and thermostats to entertainment systems and appliances using voice commands. Over time, these systems learn user preferences and automate tasks for convenience.

- **Home Security:** AI-driven **smart security cameras** and **doorbell systems** can monitor homes in **real-time**, recognize faces, and alert homeowners to potential threats. AI systems differentiate between harmless activities and real risks, providing peace of mind.

- **Energy Management:** AI-powered thermostats and appliances reduce **energy consumption** by learning the homeowner's schedule and optimizing energy use. Smart homes contribute to sustainability efforts by minimizing waste and reducing the carbon footprint.

Summary:
Smart homes, powered by artificial intelligence (AI), are transforming the way we live by offering convenience, security, energy efficiency, and even health management. From **voice assistants** like Amazon Alexa to AI-powered **security systems** and smart thermostats, **smart homes** are becoming more responsive, intuitive, and sustainable. These intelligent systems learn from user behavior, automate tasks, and help homeowners **save time, energy,** and **money** while improving their overall quality of life. As AI technology continues to evolve, the future of smart homes will see even deeper integration of ambient intelligence, making homes more autonomous and efficient while addressing challenges related to privacy and security.

028 Autonomous Drones
Revolutionizing Delivery and Transportation

Autonomous drones are rapidly transforming the fields of delivery and transportation, offering a glimpse into a future where goods and people can be moved quickly, efficiently, and with minimal human intervention. Powered by **artificial intelligence (AI)**, these drones are capable of navigating **complex environments**, making real-time decisions, and performing tasks with remarkable precision. From **delivering packages to remote areas** to revolutionizing urban **transportation**, autonomous drones are on the verge of reshaping logistics, commerce, and mobility on a global scale.

The concept of **autonomous drones** revolves around the integration of advanced **AI algorithms, computer vision,** and **GPS technology** to create flying machines that can operate independently of human control. Unlike traditional drones, which require a human pilot, autonomous drones can be programmed to complete tasks using pre-set routes or even adapt to dynamic conditions in real-time. This ability to **self-navigate** makes autonomous drones ideal for tasks such as delivering packages, surveying land, inspecting infrastructure, and even transporting people in the near future.

One of the most significant applications of autonomous drones is in the field of **delivery services**. Companies like Amazon, UPS, and DHL are investing heavily in drone delivery systems, with the goal of **speeding up** the delivery process while reducing costs and environmental impact. For instance, Amazon's Prime Air project envisions a future where drones deliver packages to customers in under 30 minutes, revolutionizing last-mile logistics. These drones can carry small to medium-sized packages, **flying directly** from distribution centers to customers' doorsteps, bypassing the traffic and delays associated with ground transportation. This not only accelerates the delivery process but also reduces the carbon footprint by eliminating the need for traditional delivery trucks.

Autonomous drones are particularly valuable in **remote and hard-to-reach areas** where traditional delivery methods are inefficient or impossible. For example, in rural or disaster-stricken regions, drones can deliver **medical supplies, vaccines,** and **emergency aid** far more quickly than vehicles can. Companies like Zipline are already using drones to deliver life-saving medical supplies to remote areas in **Africa** and **Asia**, proving that drones can be a game-changer in providing access to essential services. These drones are capable of navigating challenging terrains and weather conditions, ensuring that critical supplies reach their destinations in a timely manner.

In addition to package delivery, autonomous drones are poised to revolutionize the field of **transportation**, particularly in urban areas where traffic congestion is a growing concern. The concept of **passenger drones**—often referred to as **air taxis**—is no longer just a futuristic dream. Companies like Uber Elevate, Joby Aviation, and Volocopter are developing **electric vertical takeoff and landing (eVTOL)** aircraft that are designed to carry passengers over short distances, offering a new mode of urban transportation. These air taxis would operate **autonomously**, flying passengers above the congested streets of cities and reducing travel time significantly. With the ability to **take off and land vertically**, these drones could pick up and drop off passengers from rooftops or dedicated landing pads, making urban air mobility a practical reality.

The rise of autonomous drones in transportation doesn't stop at passenger travel. **Cargo drones** are being developed to transport larger shipments over **long distances**, providing a more efficient and cost-effective alternative to traditional ground and air freight. For example, Boeing and Airbus are working on autonomous cargo drones capable of **carrying heavy loads** across continents, which could revolutionize the global supply chain. These cargo drones would fly **autonomously**, avoiding air traffic congestion and delivering goods directly to distribution centers, factories, or retail locations. This could drastically reduce shipping times and costs while minimizing the environmental impact of traditional shipping methods.

AI plays a central role in making autonomous drones capable of **navigating complex environments** without human intervention. Using **computer vision**, drones can detect and avoid obstacles, such as buildings, trees, and other drones, in real time. **AI algorithms** allow drones to make split-second decisions based on environmental conditions, ensuring safe and efficient flight paths. In addition, AI enables drones to optimize their routes based on traffic, weather conditions, and other factors, allowing for more efficient deliveries and transportation.

The potential applications of autonomous drones extend far beyond delivery and transportation. **Agriculture** is another field that stands to benefit from drone technology. Autonomous drones are being used to **survey crops**, **monitor soil health**, and even **apply fertilizers and pesticides** with pinpoint accuracy. By analyzing data from sensors and cameras, AI-powered drones can identify areas of a field that need attention, allowing farmers to optimize their crop yields while minimizing the use of **water** and **chemicals**. Similarly, drones are being used in construction and infrastructure inspection, where they can survey large areas or inspect bridges, roads, and buildings for damage without putting human workers at risk.

However, despite the promise of autonomous drones, there are still challenges to overcome before they can be widely adopted. **Regulation** is one of the primary hurdles, as governments work to create frameworks for the safe integration of drones into airspace. **Safety concerns** and **privacy issues** also need to be addressed, as the proliferation of drones raises questions about potential accidents and the unauthorized collection of data. Additionally, the **battery life** of drones remains a limiting factor, especially for long-distance flights or heavy cargo loads, but advancements in **battery technology** and renewable energy sources are expected to improve this over time.

Despite these challenges, the **future** of **autonomous drones** is incredibly bright. As technology continues to advance, the integration of drones into everyday life will likely become more seamless, offering **faster deliveries**, **new transportation options**, and **improved services** across various industries. Governments and private companies alike are investing heavily in drone research and development, ensuring that the era of autonomous drones is not far off.

Applications:
Autonomous drones are transforming a wide range of industries, offering faster, more efficient solutions to transportation and logistics challenges. Here are some key applications:

- **Package Delivery:** Companies like Amazon and UPS are developing drone delivery systems to reduce delivery times and costs, offering **same-day** or **next-day delivery** with minimal environmental impact. Autonomous drones can deliver packages directly to customers' homes, bypassing traditional delivery methods.

- **Medical and Emergency Services:** Drones are being used to deliver **medical supplies, vaccines,** and **emergency aid** to remote areas or disaster-stricken regions. These drones are capable of reaching hard-to-access locations quickly, potentially saving lives.

- **Urban Air Mobility: Passenger drones** and **air taxis** are being developed to offer a new mode of urban transportation, reducing travel time in congested cities. These autonomous air vehicles could provide a sustainable alternative to traditional ground transportation.

Summary:
Autonomous drones, powered by AI, are revolutionizing the way we deliver goods and move people. From **package delivery** systems that offer faster, more efficient logistics to **passenger drones** that promise to reduce travel time in congested cities, drones are at the forefront of the transportation revolution. With applications ranging from **medical services** and **disaster relief** to **urban air mobility** and **cargo transport**, autonomous drones are reshaping industries across the globe. While challenges related to **regulation, safety,** and **battery life** remain, the future of autonomous drones is incredibly promising, offering new opportunities for innovation, efficiency, and sustainability in transportation.

029 Self-Healing Materials
The Next Generation of Manufacturing

Imagine a world where materials can **repair themselves**, extending the lifespan of everything from smartphones to bridges without the need for human intervention. This once futuristic concept is now becoming a **reality** with the development of **self-healing materials**—a groundbreaking innovation that is set to revolutionize the manufacturing industry. These materials can **automatically** detect and repair damage caused by wear and tear, helping reduce waste, maintenance costs, and even environmental impact. From automotive parts to construction materials and electronic devices, **self-healing technologies** represent the next generation of manufacturing, promising a future where products are not only more durable but also more sustainable.

At the core of self-healing materials is the ability to **autonomously repair** micro-damage that would otherwise lead to material failure over time. These materials are designed to respond to damage in much the same way that living organisms heal from wounds. When a crack or tear occurs, self-healing materials activate a **chemical or physical process** that restores their structure, preventing further degradation. There are several types of self-healing materials, including **polymers**, **metals**, and **composites**, each of which has its own unique method of self-repair.

One of the most common types of self-healing materials is **polymeric materials**, which are used in a wide range of applications, from consumer electronics to automotive components. These polymers contain **microcapsules** filled with **healing agents—liquids** that can flow into cracks or tears when the material is damaged. Once released, the healing agent reacts with a catalyst embedded in the material, **polymerizing** and **filling** the crack, effectively restoring the material's original strength and structure. This process occurs without any external intervention, making it ideal for environments where regular maintenance is difficult or expensive.

Self-healing composites are another class of materials with significant potential, particularly in industries like **aerospace**, **construction**, and **automotive manufacturing**. These materials are designed to self-repair damage caused by stress, fatigue, or impact, such as cracks that form in aircraft fuselages, bridge supports, or car frames. **Carbon fiber** composites, which are widely used for their strength and light weight, are now being developed with **self-healing capabilities**. By integrating **healing agents** or **nano-materials** into the composite structure, these materials can autonomously heal cracks or delamination, reducing the risk of catastrophic failure and extending the lifespan of critical infrastructure.

One of the most exciting applications of **self-healing materials** is in the field of **electronics**. As consumer devices like smartphones, tablets, and laptops become more sophisticated, they are also more prone to damage, particularly from accidental drops or wear and tear. Researchers are developing **self-healing electronics** that can repair **broken circuits** or **damaged screens**, allowing devices to recover from minor damage without needing costly repairs or replacements. For example, self-healing polymers could be integrated into **smartphone screens** that automatically repair scratches, keeping the screen smooth and clear. In the future, entire electronic systems could be built from self-healing materials, improving the longevity and reliability of devices.

Self-healing concrete is another revolutionary advancement with the potential to transform the **construction industry**. Concrete is one of the most widely used building materials in the world, but it is also prone to cracking over time due to environmental factors like temperature changes, moisture, and pressure. Cracked concrete can lead to structural weaknesses in buildings, bridges, and roads, requiring expensive repairs or even posing safety risks. To address this, scientists have developed **self-healing concrete** that incorporates **bacteria** or **chemical agents** that activate when cracks form. The bacteria or agents produce **calcium carbonate**, which fills the cracks and restores the integrity of the concrete. This self-repairing process not only reduces the need for maintenance but also extends the lifespan of infrastructure, making cities more resilient and cost-effective.

Beyond construction, **self-healing materials** are making waves in **aerospace** and **automotive** industries. Aircraft and vehicles are subjected to constant wear and stress, particularly in extreme environments. **Self-healing composites** used in aircraft wings or car body panels can repair minor cracks or damage caused by fatigue, reducing the need for frequent maintenance and lowering the overall cost of ownership. In aerospace, where material failure can have catastrophic consequences, the use of self-healing materials could improve safety and reliability while extending the service life of aircraft components.

Nanotechnology is playing a crucial role in advancing self-healing materials. **Nanomaterials** can be embedded into various types of materials to enhance their self-healing properties. For example, **nanotubes** or **nanoparticles** can be incorporated into polymers or composites to create **microvascular networks** within the material. When damage occurs, these networks deliver healing agents to the damaged area, facilitating the self-repair process. This approach not only improves the efficiency of self-healing materials but also allows them to heal multiple times, further extending their lifespan.

Despite the promise of **self-healing materials**, there are still challenges to overcome before they can be widely adopted. One of the primary challenges is **cost**. While self-healing materials have the potential to save money in the long term by reducing maintenance and extending product lifespans, the initial development and production costs can be high. Researchers are working on making self-healing materials more affordable by improving manufacturing techniques and finding cost-effective ways to integrate healing agents into materials. Additionally, the **durability** of self-healing materials is still being tested, particularly in extreme conditions such as high temperatures, heavy loads, or harsh chemical environments. Ensuring that self-healing materials can perform **reliably** in these conditions is critical for their widespread adoption.

However, the potential benefits of **self-healing materials** are vast. In a world where sustainability and resource conservation are increasingly important, materials that can repair themselves and extend their lifespan can help reduce waste and lower the demand for raw materials. This makes self-healing technologies a key component of the **circular economy**, where products are designed to last longer and be reused or recycled, reducing the environmental impact of manufacturing.

Applications:

Self-healing materials are transforming industries by offering longer-lasting, more resilient solutions. Here are some key applications:

- **Consumer Electronics:** Self-healing **polymers** in smartphones, tablets, and laptops could automatically repair scratches, cracked screens, or damaged circuits, reducing the need for costly repairs or replacements. This improves device longevity and consumer satisfaction.

- **Construction and Infrastructure:** Self-healing **concrete** and composites are revolutionizing the construction industry, with materials that can autonomously repair cracks and extend the lifespan of buildings, bridges, and roads. This reduces maintenance costs and improves safety.

- **Automotive and Aerospace:** Self-healing **composites** used in cars and airplanes can repair damage caused by stress and fatigue, reducing maintenance and improving the durability of critical components. This increases reliability and safety, particularly in extreme environments.

Summary:

Self-healing materials are revolutionizing manufacturing by enabling products and infrastructure to repair themselves, reducing waste, extending lifespans, and improving sustainability. From self-healing **polymers** in electronics to self-repairing **composites** in the automotive and aerospace industries, these materials are designed to automatically respond to **damage** and restore their structural integrity. In construction, self-healing **concrete** is set to transform how we build cities, reducing the need for **costly repairs** and improving the **longevity** of infrastructure. While challenges related to **cost** and **durability** remain, the future of self-healing materials is incredibly promising, offering a new generation of products that are not only more durable but also more environmentally friendly.

030 Exoskeletons
Enhancing Human Mobility

Exoskeletons, once the stuff of science fiction, are now becoming a reality, offering new possibilities for enhancing **human mobility** and strength. These wearable devices, designed to augment physical abilities, are transforming the lives of individuals with **disabilities**, revolutionizing the way people work in industries requiring physical labor, and opening up new frontiers in **rehabilitation** and **healthcare**. Powered by **advanced robotics** and **artificial intelligence (AI)**, exoskeletons are poised to reshape how we think about human potential by providing assistance for those who need it and enhancing the capabilities of those looking to push their physical limits.

At their core, **exoskeletons** are wearable machines that fit over the body and assist with movement, strength, or endurance. By using **motors, actuators**, and **AI-driven sensors**, exoskeletons can detect a user's movements and provide the necessary support to enhance strength, mobility, or stability. These devices can be used in a wide range of applications, from **healthcare** and **rehabilitation** to **industrial work** and even **military applications**. Whether helping a **paraplegic** walk again or reducing the strain on factory workers lifting heavy objects, exoskeletons represent a major leap forward in wearable technology.

One of the most impactful uses of exoskeletons is in the field of **rehabilitation** and **assistive technology**. For individuals with **spinal cord injuries, stroke**, or other conditions that limit mobility, exoskeletons provide a way to regain the ability to stand, walk, and perform daily tasks. Devices like **ReWalk, Ekso Bionics**, and **Indego** are some of the leading exoskeletons designed to help people with lower limb paralysis regain mobility. These devices use **sensors** and **motors** to detect a user's intention to move and provide the necessary mechanical assistance to enable walking. By supporting the user's weight and guiding their movement, exoskeletons allow individuals to stand upright, walk, and even climb stairs—activities that may have been impossible for them due to injury or illness.

The impact of exoskeletons on **rehabilitation** is profound. Studies have shown that using exoskeletons for **gait training** can significantly improve mobility and quality of life for individuals recovering from a stroke or traumatic injury. By enabling users to practice walking in a controlled and supported manner, exoskeletons help **retrain muscles**, improve **balance**, and promote **neuroplasticity**—the brain's ability to reorganize itself after injury. This means that, over time, users may regain some natural mobility even without the exoskeleton, making these devices a critical tool in modern physical therapy.

Exoskeletons are also making their way into the workplace, particularly in industries that require **physical labor**. In sectors like **manufacturing**, **construction**, and **warehousing**, workers are often required to lift heavy objects, bend repeatedly, or perform physically demanding tasks for extended periods. These activities can lead to fatigue, injuries, and long-term musculoskeletal problems. **Industrial exoskeletons** are designed to reduce the strain on workers by providing mechanical assistance with lifting, supporting the back, or reinforcing the legs. Companies like SuitX, Ottobock, and Sarcos Robotics have developed exoskeletons that can help workers lift heavy objects with ease, reducing the risk of injury and improving productivity. By distributing the load more evenly across the body and providing extra strength, these exoskeletons enable workers to perform their tasks with less effort and greater safety.

In addition to enhancing **physical strength**, exoskeletons are also being used to improve **endurance** and **stamina**. For example, **military exoskeletons** are being developed to help soldiers carry heavy equipment over long distances without becoming fatigued. These exoskeletons, like the Lockheed Martin FORTIS, reduce the physical burden on soldiers by supporting their bodies and providing mechanical assistance for lifting and carrying loads. This not only improves the soldiers' endurance but also reduces the risk of injury during long missions or strenuous activities. The **military** sees exoskeleton technology as a way to enhance the capabilities of its personnel, making them more efficient and resilient in the field.

One of the most promising advancements in exoskeleton technology is the integration of **AI** and **machine learning**. By incorporating AI-driven **sensors** and **algorithms**, exoskeletons can learn from a user's movements and adapt to their specific needs. For example, AI can analyze gait patterns, balance, and strength to provide personalized assistance that is tailored to the user's abilities. This allows for more natural and intuitive movement, making exoskeletons easier to use and more effective in enhancing mobility. In the workplace, AI-driven exoskeletons can optimize assistance based on the tasks being performed, improving efficiency and reducing fatigue for workers.

Medical exoskeletons are also being developed for individuals with degenerative conditions like **muscular dystrophy** or **multiple sclerosis (MS)**. These devices provide ongoing support as the user's muscles weaken, enabling them to maintain mobility and independence for longer periods. In some cases, exoskeletons can be used in conjunction with other assistive technologies, such as **brain-computer interfaces (BCIs)**, allowing users to control their movements through thought alone. This combination of exoskeletons and BCIs represents the cutting edge of assistive technology, offering new hope to individuals with severe disabilities.

Despite their potential, there are still challenges to overcome in the widespread adoption of exoskeletons. One of the primary challenges is **cost**. Exoskeletons are complex devices that require **advanced materials**, **sensors**, and **robotics**, making them expensive to produce and purchase. While prices are expected to come down as the technology matures and scales, cost remains a barrier for many individuals and organizations. Another challenge is ensuring that exoskeletons are **comfortable** and **easy to use** for extended periods. Engineers are working to make these devices lighter, more flexible, and less cumbersome, so they can be worn for long durations without causing discomfort or fatigue.

Applications:

Exoskeletons are transforming the way we think about human mobility and physical capability. Here are some of the key applications:

- **Rehabilitation and Assistive Technology**: Exoskeletons like **ReWalk** and **Ekso Bionics** are helping individuals with paralysis or mobility impairments regain the ability to walk and perform daily tasks. These devices are also used in **gait training** for stroke recovery, promoting **muscle retraining** and **neuroplasticity**.

- **Industrial Work:** In industries like **construction** and **manufacturing**, exoskeletons reduce the physical strain on workers by assisting with lifting and repetitive tasks. This helps prevent injuries, improves worker efficiency, and enhances safety on the job.

- **Military and Defense:** Exoskeletons are being developed to help soldiers carry **heavy equipment** and improve **endurance** during long missions. These devices reduce fatigue and lower the risk of injury, allowing soldiers to perform more effectively in the field.

Summary:

Exoskeletons are revolutionizing **human mobility** by providing mechanical support to individuals with disabilities, enhancing the physical capabilities of workers, and improving endurance in military personnel. From **rehabilitation devices** that allow paraplegics to walk again to **industrial exoskeletons** that reduce strain on workers, these wearable machines are transforming industries and improving quality of life. With advancements in **AI** and **machine learning**, exoskeletons are becoming smarter, more adaptable, and easier to use, offering personalized support that enhances natural movement. As costs come down and technology improves, exoskeletons are set to play a key role in the future of **healthcare**, **workplace safety**, and **human augmentation**.

031 Deep-Sea Exploration
Unlocking the Secrets of the Ocean

Deep-sea exploration represents one of the most exciting frontiers of science and discovery. Covering more than **70% of the Earth's surface**, the oceans remain largely unexplored, with only a small fraction of the seafloor having been mapped or studied. The **deep sea** holds untold mysteries, from unique ecosystems and undiscovered species to resources that could revolutionize industries. With advancements in **robotics**, **submersible technology**, and **artificial intelligence** (**AI**), scientists are now able to delve deeper into the ocean than ever before, unlocking the secrets of the **abyss** and shedding light on a world that has been hidden from view for millennia.

The **deep sea** begins at depths of **200 meters and beyond**, where sunlight no longer penetrates and extreme conditions like high pressure, frigid temperatures, and complete darkness dominate. The technology required to explore these depths must be robust enough to withstand these harsh environments. In recent years, **autonomous underwater vehicles (AUVs)** and **remotely operated vehicles (ROVs)** have become essential tools for deep-sea exploration. These highly advanced machines are equipped with **high-definition cameras, sonar, and sensors** that can capture data and images from the deepest parts of the ocean. AUVs can operate autonomously, navigating the ocean floor and collecting samples, while ROVs allow scientists to control exploration missions from the surface, sending back real-time footage of their discoveries.

One of the most thrilling aspects of deep-sea exploration is the discovery of **new species**. The deep ocean is home to some of the most bizarre and alien creatures on the planet, many of which are uniquely adapted to survive in extreme conditions. Animals like the **giant squid**, the **anglerfish**, and the **vampire squid** have evolved extraordinary adaptations such as **bioluminescence**, specialized feeding mechanisms, and the ability to withstand crushing pressures. With each new expedition, scientists are uncovering species that were previously unknown to science, helping to expand our understanding of life on Earth. In 2020, for example, an expedition led by **Schmidt Ocean Institute** discovered over **30 new species** in the **deep waters of the Indian Ocean**, revealing the incredible biodiversity that exists in the ocean's depths.

Beyond biological discoveries, deep-sea exploration is also uncovering **geological phenomena** that challenge our understanding of Earth's processes. One of the most important discoveries of the past few decades has been **hydrothermal vents**—underwater geysers that release superheated, mineral-rich water from beneath the Earth's crust. These vents, found along **mid-ocean ridges**, are home to unique ecosystems that thrive in complete darkness, powered not by sunlight but by **chemosynthesis**, where organisms use chemicals from the vent fluid to produce energy. The discovery of these ecosystems has reshaped our understanding of the limits of life on Earth and even holds implications for the search for life on other planets, particularly in the subsurface oceans of moons like **Europa** and **Enceladus**.

Deep-sea exploration is also unlocking **mineral resources** that could revolutionize industries. The ocean floor is rich in deposits of **polymetallic nodules, manganese crusts**, and **hydrothermal sulfide deposits** that contain valuable metals like **copper, nickel, cobalt**, and **rare earth elements**—materials that are essential for modern technologies such as **batteries, smartphones**, and **renewable energy systems**. Mining these resources could provide a new supply of critical materials, reducing dependence on terrestrial mining and helping to meet the demands of the **green energy** transition. Companies and governments are actively exploring the potential for **deep-sea mining**, though concerns about the environmental impact are leading to debates about how to balance resource extraction with the need to protect fragile marine ecosystems.

One of the most fascinating frontiers of deep-sea exploration is the search for **submarine volcanoes** and **seamounts**—underwater mountains that rise from the ocean floor. These geological features are often teeming with life, serving as important habitats for marine species and as **hotspots** for biodiversity. Submarine volcanoes are of particular interest because they can provide insights into **plate tectonics, earthquakes**, and the formation of new oceanic crust. The study of these underwater volcanoes also offers clues about Earth's geological history and the dynamic processes that continue to shape the planet.

As the technology for deep-sea exploration continues to advance, AI is playing an increasingly important role in processing the vast amounts of data collected during ocean expeditions. **AI algorithms** are being used to **analyze sonar data**, identify patterns in marine life behavior, and even map the ocean floor with unprecedented detail. This technology is enabling scientists to make faster and more accurate discoveries, helping to unlock the mysteries of the deep ocean in ways that were previously impossible. By automating data analysis and integrating machine learning, AI is helping researchers uncover new insights about ocean ecosystems, geology, and even climate change.

Deep-sea exploration also holds the key to understanding the **ocean's role in regulating climate**. The oceans absorb vast amounts of **carbon dioxide** from the atmosphere, acting as a critical buffer against climate change. Studying the deep sea helps scientists understand how carbon is stored and cycled through the ocean's depths, providing valuable information for climate models and strategies to mitigate global warming. The discovery of deep-sea ecosystems that act as carbon sinks, such as **cold seeps**, is helping researchers understand the complex interactions between the ocean and the atmosphere, offering new opportunities for combating climate change.

Applications:
Deep-sea exploration is leading to groundbreaking discoveries that impact a wide range of fields. Here are some key applications:

- **Biological Discoveries:** Exploring the deep ocean is uncovering new species and ecosystems, expanding our understanding of life on Earth. These discoveries hold potential for medical and biotechnological innovations.

- **Geological Insights:** Studying **hydrothermal vents, submarine volcanoes**, and the ocean floor is providing valuable information about **plate tectonics, earthquakes**, and the formation of new geological features.

- **Resource Exploration:** The deep sea is rich in valuable minerals like **copper**, **nickel**, and **rare earth elements**, offering new opportunities for sustainable resource extraction.

Summary:
Deep-sea exploration is unlocking the secrets of the ocean's depths, revealing new species, ecosystems, and geological phenomena that are expanding our understanding of the planet. From the discovery of **hydrothermal vents** and unique **deep-sea creatures** to the exploration of valuable **mineral resources** and **submarine volcanoes**, advancements in **robotics, AI**, and **submersible technology** are making it possible to explore the vast and mysterious world beneath the waves. These discoveries are not only reshaping our understanding of Earth's natural systems but also providing critical insights into the ocean's role in **climate regulation** and offering potential solutions for sustainable resource management. The future of deep-sea exploration promises even more exciting breakthroughs as we continue to uncover the hidden wonders of the ocean.

032 Advanced AI
Autonomous Warfare

The rise of **advanced artificial intelligence (AI)** is reshaping the landscape of **warfare**, introducing new possibilities and challenges as nations race to develop autonomous systems capable of executing military operations with minimal human intervention. From **drones** and **robotic soldiers** to AI-driven **cyber defense systems**, the integration of AI into military strategy is transforming the way wars are fought. As AI systems become more sophisticated, they are not only enhancing the capabilities of armed forces but also raising important ethical questions about the future of **autonomous warfare**. The rapid development of AI in the military domain signals a new era in defense, where speed, precision, and adaptability are critical.

At the forefront of this revolution are **autonomous drones**, which have become key players in modern military operations. These drones, equipped with advanced AI systems, are capable of carrying out missions such as **surveillance**, **reconnaissance**, and **targeted strikes** with minimal or no human oversight. Drones like the MQ-9 Reaper and Bayraktar TB2 are examples of how AI is enhancing the **precision** and **efficiency** of military engagements. These systems use AI algorithms to process vast amounts of **data** from **sensors** and **cameras**, allowing them to identify targets, avoid obstacles, and adjust their flight paths in real time. By automating these tasks, AI enables drones to perform complex missions faster and more accurately than human pilots could.

One of the most significant advantages of AI in **autonomous warfare** is its ability to **process** and **analyze data** at incredible speeds. On the **battlefield**, timely and accurate information is critical to success. AI-driven systems can analyze data from multiple sources, such as **satellite imagery, radar**, and **reconnaissance drones**, to provide real-time insights into enemy movements and terrain conditions. This allows commanders to make informed decisions more quickly, gaining a tactical advantage over adversaries. Additionally, AI-powered systems can **predict enemy behavior** by analyzing historical data and identifying patterns, helping military strategists anticipate attacks or disruptions before they occur.

AI in autonomous warfare is also revolutionizing **cyber defense** and **cyber warfare**. With modern warfare increasingly relying on digital infrastructure, protecting military networks and systems from cyberattacks has become a top priority. AI systems are being developed to detect and respond to cyber threats in real time, using **machine learning algorithms** to identify vulnerabilities and neutralize attacks before they cause damage. These AI-driven cyber defense systems can monitor vast networks for unusual activity, recognize patterns associated with malicious behavior, and **automatically** respond to threats without waiting for human intervention. This rapid response capability is essential in defending against increasingly sophisticated cyberattacks launched by state and non-state actors.

In addition to **cyber defense**, AI is enhancing the capabilities of **robotic soldiers** and **autonomous ground vehicles**. These systems are being designed to operate in environments that are too dangerous for human soldiers, such as urban combat zones or battlefields contaminated with chemical or biological agents. **AI-powered robots** can perform a variety of tasks, from clearing mines and transporting supplies to engaging in combat alongside human troops. These robots are equipped with AI that allows them to navigate complex environments, identify threats, and respond to changing conditions in real time. By reducing the need for human soldiers in high-risk scenarios, AI-driven robots can **minimize casualties** while increasing operational effectiveness.

One of the most cutting-edge developments in autonomous warfare is the use of **AI swarms**—groups of drones or robots that work together as a coordinated unit. These swarms can be deployed to overwhelm enemy defenses, conduct reconnaissance missions, or provide logistical support in dangerous areas. Using AI algorithms, the drones in a swarm can **communicate** with each other, share **information**, and adapt their **strategies** based on the situation at hand. The potential of AI swarms in warfare is immense, as they offer a scalable and flexible approach to military operations, capable of quickly adapting to changing battlefield conditions. Military researchers are exploring how to use these swarms for a range of missions, from **search-and-rescue** operations to **targeted strikes**.

As AI becomes more integrated into military systems, the ethical implications of **autonomous warfare** are coming under increasing scrutiny. One of the key concerns is the role of human decision-making in the use of **lethal force**. Fully autonomous systems, often referred to as **lethal autonomous weapons systems (LAWS)**, raise questions about accountability and control. Should machines be allowed to make life-and-death decisions on their own? Military leaders and ethicists are debating how to balance the advantages of autonomous systems with the need to ensure that human oversight remains a key component of warfare. Nations around the world are working to develop frameworks that address these ethical concerns, with some calling for **international regulations** on the use of autonomous weapons.

The development of **AI in warfare** is also leading to an arms race among global powers. Countries like the United States, China, and Russia are investing heavily in AI research and development, seeking to gain a technological edge in military capabilities. This **competition** is driving innovation in **autonomous systems**, from **drones** and **robotic soldiers** to **cyber defense** and **AI-driven logistics**. As nations strive to outpace one another in the development of military AI, the potential for these technologies to reshape global security is becoming increasingly clear. Military planners are now factoring AI into their strategies, recognizing that future conflicts may be won or lost based on the capabilities of autonomous systems.

While the potential benefits of **AI in autonomous warfare** are vast, the technology also presents significant challenges. The rapid pace of AI development means that military systems are becoming more complex, requiring advanced skills to manage and maintain. Additionally, ensuring the security of AI-driven systems is a top concern, as adversaries may attempt to **hack** or **manipulate** autonomous weapons to turn them against their operators. Developing robust safeguards and **fail-safe mechanisms** to prevent unauthorized control or misuse of these systems is critical to their success.

Applications:

The integration of AI in autonomous warfare is revolutionizing military operations across multiple domains. Here are some key applications:

- **Autonomous Drones:** AI-powered drones are being used for **surveillance, reconnaissance**, and **targeted strikes**. These drones can operate independently, navigating complex environments and identifying targets in real time.

- **Cyber Defense Systems:** AI-driven cyber defense systems monitor military networks for cyber threats, responding to potential attacks before they cause damage. AI is enhancing the ability to defend against increasingly sophisticated cyberattacks.

- **AI Swarms:** AI swarms of drones or robots can work together to conduct missions, providing scalable and adaptive strategies in combat or search-and-rescue operations.

Summary:

Advanced AI in autonomous warfare is transforming military operations, offering new capabilities in **surveillance, cyber defense,** and **combat systems**. From **autonomous drones** capable of executing missions without human intervention to **AI-powered cyber defense systems** that respond to threats in real time, AI is enhancing the speed, precision, and adaptability of military forces. As nations invest heavily in developing **robotic soldiers** and **AI swarms**, the future of warfare is shifting toward a world where autonomous systems play a central role in both offensive and defensive operations. While the technology offers significant advantages, ethical concerns about the use of **lethal autonomous weapons systems (LAWS)** and the potential for misuse remain critical issues that must be addressed. The race to develop AI-driven warfare capabilities is shaping the future of global security and military strategy.

033 Bioprinting
Printing Human Tissues and Organs

Bioprinting is a revolutionary technology that is transforming the field of medicine by making it possible to print **human tissues and organs**. Using techniques similar to **3D printing**, bioprinting involves the precise layering of **biomaterials**, including **cells, growth factors**, and **biocompatible polymers**, to create structures that mimic the form and function of natural tissues. This cutting-edge technology has the potential to solve one of the most critical challenges in healthcare—finding replacement organs for patients in need. With advancements in bioprinting, the future of **organ transplantation** and **tissue engineering** is being reimagined, offering hope to millions of patients waiting for life-saving treatments.

At its core, **bioprinting** uses a 3D printer to layer **bio-inks**, which are composed of living cells, to build up tissue structures one layer at a time. These bio-inks can be customized to replicate the different types of **cells** found in the human body, such as **muscle cells, nerve cells**, or **skin cells**. By precisely controlling the placement of these cells, bioprinting can create complex tissue architectures that closely resemble the real thing. The process begins with a **computer-generated model** of the tissue or organ to be printed, which guides the printer in building the structure layer by layer.

One of the most **promising** applications of bioprinting is in the field of **regenerative medicine**, where damaged or diseased tissues can be replaced with **lab-grown alternatives**. Researchers are already using bioprinting to create functional tissues such as **skin, bone**, and **cartilage** for use in medical treatments. For example, bioprinted skin grafts can be used to treat **burn victims**, providing a more effective and natural-looking solution than traditional skin grafts. Similarly, bioprinted cartilage can be used to repair damaged joints, offering hope to patients suffering from **arthritis** or sports injuries. As the technology advances, scientists are working toward the ultimate goal of printing **functional human organs**, such as **kidneys**, **livers**, and **hearts**, which could one day be transplanted into patients in need.

One of the key challenges in **organ transplantation** is the severe shortage of donor organs. Thousands of patients die each year waiting for a suitable organ match, as the demand for organs far exceeds the supply. **Bioprinting** offers a potential solution to this crisis by enabling the creation of **personalized organs** that are tailor-made for each patient. By using the patient's own cells to bioprint the organ, doctors can reduce the risk of **organ rejection**, which is a common complication in traditional transplants. This personalized approach to organ creation could revolutionize the field of transplantation, providing patients with **custom-made organs** that are fully compatible with their bodies.

Bioprinting is also opening up new possibilities in **drug development** and **medical research**. Traditionally, new drugs are tested on animals or human cell cultures, which do not always accurately replicate the complexity of human tissues. With bioprinted tissues, pharmaceutical companies can test new drugs on **lab-grown human tissues**, providing more accurate data on how the drugs will interact with the human body. This approach not only speeds up the drug development process but also reduces the need for animal testing. In addition, bioprinting can be used to create **disease models**—tissues that mimic the characteristics of specific diseases—allowing researchers to study the progression of diseases like **cancer**, **Alzheimer's**, and **heart disease** in a controlled environment.

The development of bioprinted **vascular tissues** is one of the most critical advancements in the field. For bioprinted organs to be functional, they need to be able to receive **oxygen** and **nutrients** through a network of **blood vessels**. Researchers are making significant progress in printing **vascularized tissues**, which include networks of tiny blood vessels capable of transporting blood throughout the tissue. This breakthrough is essential for the creation of large, **complex organs**, such as kidneys and hearts, which require a functioning vascular system to survive and integrate into the body.

Bioprinting also holds promise for the field of **cosmetic and reconstructive surgery**. Patients who have suffered from traumatic injuries, congenital deformities, or conditions like **cancer** often require **reconstructive surgery** to restore function and appearance. Bioprinting can be used to create customized **implants** and **prosthetics** that are more natural and better suited to the patient's anatomy than traditional implants. For example, bioprinted **bone grafts** can be used to replace damaged or missing bone tissue, while bioprinted ear **cartilage** can be used to reconstruct ears in patients with congenital deformities. This technology allows surgeons to create highly personalized solutions that improve patient outcomes and quality of life.

Another exciting development in the field is the bioprinting of **miniature organs**, also known as organoids. These small, simplified versions of human organs are grown from stem cells and can be used for a wide range of applications, from drug testing to studying the development of diseases. **Organoids** provide a valuable tool for understanding complex biological processes and testing the effects of new treatments on specific organs, such as the liver or kidneys. In the future, bioprinted organoids could be used to study the effects of genetic mutations, test personalized therapies for individual patients, and even develop new treatments for **rare diseases**.

The integration of **AI** and **machine learning** into the bioprinting process is helping researchers improve the precision and efficiency of bioprinted tissues. AI algorithms can analyze large datasets and optimize the printing parameters for different types of tissues, ensuring that the printed structures are as close to natural tissues as possible. AI can also be used to simulate how bioprinted tissues will behave once implanted in the body, allowing scientists to make adjustments before the tissue is printed. This integration of AI is accelerating the development of bioprinting and bringing the technology closer to **clinical applications**.

While bioprinting has made significant strides in recent years, there are still **challenges** that need to be addressed before the technology can be widely used in clinical settings. One of the main challenges is ensuring the **long-term viability** of bioprinted tissues and organs once they are implanted in the body. Researchers are working to develop **better materials** and **techniques** that can support the growth and function of bioprinted tissues over time. Additionally, the cost of bioprinting remains high, and further advancements are needed to make the technology more affordable and accessible.

Applications:
Bioprinting is revolutionizing several fields of medicine and research. Here are some key applications:

- **Organ Transplantation:** Bioprinted organs can be customized for each patient, reducing the risk of rejection and providing a solution to the shortage of donor organs. This personalized approach could transform the field of transplantation.

- **Tissue Engineering and Regenerative Medicine:** Bioprinted **tissues**, such as **skin**, **bone**, and **cartilage**, are being used to treat injuries, burns, and degenerative diseases, providing patients with functional and natural-looking replacements.

- **Drug Testing and Disease Modeling:** Bioprinted tissues are used to test new drugs and study the progression of diseases like **cancer** and **heart disease**, improving the accuracy of medical research and reducing the need for animal testing.

Summary:
Bioprinting is revolutionizing healthcare by making it possible to print **human tissues** and **organs**, offering new solutions for **transplantation, tissue engineering**, and **medical research**. From bioprinted **skin grafts** for burn victims to the development of **personalized organs** for transplant patients, the ability to create living tissues using **3D printing technology** is transforming medicine. This technology is also advancing **drug development** and **disease research** by allowing scientists to test treatments on lab-grown human tissues. As progress continues, bioprinting holds the potential to address the critical shortage of donor organs and improve patient outcomes across a wide range of medical fields.

034 Carbon Nanotubes
The Strongest Material Known to Man

Carbon nanotubes (CNTs) are widely regarded as one of the most remarkable discoveries in **material science**. These cylindrical structures, made entirely of **carbon atoms** arranged in a honeycomb lattice, possess extraordinary properties that have the potential to revolutionize industries from **electronics** and **energy** to **construction** and **medicine**. Carbon nanotubes are incredibly strong, lightweight, and conductive, making them the **strongest material known to man**. With applications ranging from ultra-durable materials to **advanced electronics**, CNTs are at the forefront of nanotechnology, offering unprecedented possibilities for innovation and design.

At their core, **carbon nanotubes** are a unique form of carbon. Their structure is essentially a single layer of **graphene**—a two-dimensional sheet of carbon atoms—rolled into a cylindrical tube. These tubes can be **single-walled**, with a diameter of just a few nanometers, or **multi-walled**, with several concentric graphene tubes nested within each other. The incredible strength of CNTs comes from the strong **carbon-carbon bonds** that hold the atoms together, making them more than **100 times stronger than steel** at a fraction of the weight. This combination of strength and lightness has led to CNTs being hailed as a potential game-changer in industries that require high-performance materials.

One of the most **promising applications** of **carbon nanotubes** is in the field of **aerospace** and **automotive manufacturing**. The need for **lightweight** yet **durable** materials is critical in these industries, where reducing weight can lead to significant improvements in fuel efficiency and performance. CNTs are being integrated into **composites** that can be used to build aircraft frames, spacecraft components, and car parts that are both lighter and stronger than traditional materials. By using **CNT-reinforced composites**, manufacturers can improve the overall performance and safety of vehicles while reducing energy consumption and emissions.

In addition to their mechanical strength, **carbon nanotubes** are excellent **conductors of electricity** and **heat**, making them valuable in the **electronics** industry. CNTs have the potential to replace traditional materials like **silicon** in electronic components, enabling the creation of smaller, faster, and more efficient devices. One of the most exciting possibilities is the development of **CNT-based transistors**, which could allow for the creation of faster and more powerful computer chips. As the demand for more compact and efficient electronics continues to grow, CNTs offer a solution that can keep pace with the increasing need for higher performance in smaller form factors.

The unique electrical properties of **carbon nanotubes** also make them ideal for **energy storage** applications. CNTs can be used to create **supercapacitors** and **batteries** with higher energy density and faster charge/discharge rates than current technologies. This could have a profound impact on industries like **renewable energy** and **electric vehicles**, where energy storage is a critical component of performance. For example, **CNT-enhanced batteries** could extend the range of electric vehicles, reduce charging times, and improve the overall efficiency of energy storage systems. Similarly, **CNT-based supercapacitors** could provide faster energy delivery for applications requiring high power output, such as grid stabilization and renewable energy integration.

Carbon nanotubes are also being explored for their potential in **medicine**. Due to their **biocompatibility** and **high surface area**, CNTs are being used as **drug delivery systems** that can target specific areas of the body with precision. Researchers are developing **CNT-based drug carriers** that can transport therapeutic agents directly to cancer cells, reducing side effects and improving treatment efficacy. The high surface area of CNTs allows for a greater concentration of drugs to be delivered, making treatments more effective. Additionally, CNTs are being used in **biosensors** that can detect diseases at an early stage by identifying specific biomarkers in the body. This technology has the potential to revolutionize diagnostics, enabling faster and more accurate detection of diseases like cancer, diabetes, and infectious diseases.

Another area where **carbon nanotubes** are making an impact is in the field of **construction**. The incredible strength and lightness of CNTs make them ideal for creating **ultra-strong building materials** that can withstand extreme forces while remaining lightweight. **CNT-reinforced concrete** is one such material that is being developed for use in the construction of **bridges, skyscrapers**, and **infrastructure** that requires both strength and flexibility. By integrating CNTs into concrete, researchers are able to improve the material's resistance to cracking, increase its durability, and extend its lifespan, reducing the need for repairs and maintenance over time.

The use of **carbon nanotubes** in **water filtration** and **environmental cleanup** is another promising application. CNTs are capable of filtering out contaminants at the molecular level, making them highly effective at purifying water. In particular, **CNT-based filters** can remove **heavy metals**, **organic pollutants**, and **bacteria** from water sources, providing a solution for regions facing water scarcity or contamination. Additionally, CNTs are being explored for their ability to capture and store **carbon dioxide (CO_2)**, which could help mitigate the effects of climate change by reducing the amount of CO_2 in the atmosphere.

One of the key challenges in the widespread adoption of **carbon nanotubes** is the difficulty of **scaling production**. While CNTs can be produced in small quantities with high precision, scaling up the manufacturing process to produce **large volumes** at an affordable **cost** remains a significant hurdle. Researchers are working on developing more efficient methods for producing CNTs in bulk, which will be critical for their widespread use in industries like **construction, electronics**, and **energy storage**. Once large-scale production is achieved, the potential of CNTs to revolutionize multiple industries will become even more apparent.

Another challenge is ensuring the **safety** and **sustainability** of CNTs, particularly in applications where they may come into contact with humans or the environment. Studies are being conducted to assess the potential risks of CNT exposure, particularly in areas like medicine and environmental cleanup. Ensuring that CNTs are **safe** for widespread use is essential to their successful integration into various industries.

Applications:
Carbon nanotubes are transforming industries by offering unparalleled strength, conductivity, and versatility. Here are some key applications:

- **Aerospace and Automotive Manufacturing:** CNT-reinforced composites are used to create lightweight, ultra-strong materials for aircraft, spacecraft, and vehicles, improving performance and fuel efficiency.

- **Electronics:** CNTs have the potential to replace traditional materials like **silicon** in transistors, creating faster, smaller, and more efficient electronic devices. CNTs are also being used to develop advanced energy storage solutions for **batteries** and **supercapacitors**.
- **Medicine:** CNTs are being explored for use in **drug delivery systems** and **biosensors**, offering new ways to target diseases and improve diagnostics with precision and efficiency.

Summary:

Carbon nanotubes (CNTs) are revolutionizing industries with their unmatched strength, lightweight properties, and exceptional conductivity. From aerospace and automotive manufacturing to electronics and medicine, CNTs offer a wide range of applications that can improve performance, efficiency, and sustainability. CNT-based composites provide stronger and lighter materials for vehicles, while CNTs' electrical properties enable the development of faster, more efficient electronic components. In medicine, CNTs are advancing drug delivery and disease detection technologies. As production methods improve, carbon nanotubes are set to play a transformative role in the future of **nanotechnology** and industry.

035 Next-Gen 5G Networks
Connecting the Future

The arrival of **next-generation 5G networks** is set to revolutionize the way we connect and interact with the world around us. Offering **faster speeds, lower latency,** and **increased connectivity**, 5G is much more than just an upgrade from previous wireless networks. It represents a fundamental shift in how data is transmitted, enabling everything from **smart cities** and **autonomous vehicles** to **virtual reality (VR)** and **augmented reality (AR)** applications. With the potential to power emerging technologies and industries, 5G is unlocking new possibilities that will shape the future of communication, commerce, and innovation.

At its core, 5G delivers **speeds up to 100 times faster** than 4G, with peak download rates reaching 10 Gbps. This speed increase means users can download movies, apps, and other large files in seconds, but the real impact of 5G lies in its ability to support **ultra-low latency** and **massive device connectivity**. Latency—the delay between sending and receiving data—is reduced to as little as **1 millisecond** with 5G, making it ideal for applications that require **real-time interaction**, such as **remote surgery**, **robotics**, and **self-driving cars**. Additionally, 5G can connect millions of devices per square kilometer, enabling the rise of the **Internet of Things (IoT)** and the development of smart homes, smart cities, and more efficient industries.

One of the most significant advantages of **5G networks** is their ability to support autonomous vehicles. The development of **self-driving cars** requires **constant communication** between the vehicle and its surroundings—other vehicles, traffic signals, and infrastructure. With 5G, autonomous cars can transmit and receive massive amounts of data in real-time, ensuring the safety and efficiency of **autonomous driving**. This communication allows vehicles to anticipate potential hazards, avoid accidents, and respond instantly to changes in traffic conditions. The ultra-low latency of 5G is essential for this, as even a slight delay in data transmission could compromise safety. By enabling **vehicle-to-vehicle (V2V)** and **vehicle-to-infrastructure (V2I)** communication, 5G is laying the groundwork for the future of transportation.

Another exciting application of 5G is in the field of **virtual reality (VR)** and **augmented reality (AR)**. These immersive technologies require significant bandwidth and low latency to function smoothly, which has been a limitation of previous networks. With 5G, users can experience seamless VR and AR interactions without the lag or buffering issues that can disrupt the immersive experience. This opens up new possibilities for **gaming, education, healthcare**, and **remote work**. For example, medical professionals can use VR to practice **complex surgeries in real time**, or students can participate in **virtual field trips** that take them to different parts of the world. The integration of 5G with AR will also enhance industries like retail and manufacturing, where AR applications can improve productivity and customer experience.

5G is also set to transform industries through the rise of the **Industrial Internet of Things (IIoT)**. In manufacturing and logistics, 5G will enable factories and warehouses to operate more efficiently by connecting robots, sensors, and equipment in real time. This will lead to **smart factories**, where machines can communicate with each other, anticipate maintenance needs, and adjust production processes without human intervention. In agriculture, **5G-powered IoT** devices can monitor crop conditions, automate irrigation systems, and optimize harvests based on real-time data. These advancements are paving the way for a new era of automation, where industries can reduce costs, improve efficiency, and make data-driven decisions with unprecedented accuracy.

Smart cities are another area where 5G will play a transformative role. By enabling **IoT networks** that connect everything from **traffic lights** and **surveillance cameras** to **public transportation** and **waste management** systems, 5G will make cities more efficient and sustainable. Traffic management systems, for instance, can be optimized with real-time data to reduce congestion, improve air quality, and enhance public safety. **Energy grids** can become more efficient by dynamically adjusting to usage patterns, while **smart streetlights** can dim or brighten based on environmental conditions. The result is a more connected, efficient, and responsive urban environment that improves the quality of life for residents.

In **healthcare**, the potential of 5G networks is vast. **Telemedicine** is one area where 5G will have a significant impact, allowing doctors to remotely monitor patients, conduct virtual consultations, and even perform surgeries with **robotic assistance**. The low latency and high reliability of 5G make it possible to provide real-time care to patients, no matter where they are located. Wearable health devices, which collect data on vital signs and patient activity, will be able to transmit **information instantly** to healthcare providers, enabling more proactive and personalized care. This is especially important for patients with chronic conditions who require **continuous monitoring** and treatment adjustments.

5G is also driving advancements in **smart home technology**, making homes more connected and responsive to residents' needs. **Smart home devices**, such as thermostats, lighting systems, security cameras, and voice assistants, can communicate seamlessly with one another and adjust settings automatically based on user preferences. For instance, a **smart thermostat** can optimize heating and cooling by learning the habits of the household, while **smart lighting systems** can adjust brightness based on natural light levels.
These connected devices, powered by 5G, create a more comfortable, energy-efficient living environment.

As 5G networks continue to expand, AI will play an integral role in managing the massive amounts of data generated by connected devices. **AI algorithms** can analyze this data in real-time to provide actionable insights, optimize network performance, and enhance security. **AI-driven analytics** will help businesses make data-driven decisions faster, improve customer experiences, and predict equipment failures before they happen. In industries like retail, AI can analyze consumer behavior data collected from **5G-enabled devices** to create personalized shopping experiences and improve inventory management.

5G is also paving the way for advancements in **public safety** and **emergency response**. With real-time data from surveillance cameras, drones, and sensors, law enforcement agencies and first responders can respond more quickly to incidents and make informed decisions in crisis situations. For example, during a natural disaster, **5G-connected drones** can be deployed to survey the area, locate survivors, and deliver supplies. Emergency response vehicles equipped with **5G connectivity** can receive real-time updates on traffic and route conditions, ensuring that help arrives as quickly as possible.

While 5G is still in the process of being rolled out **globally**, the technology is rapidly gaining momentum as telecom providers, governments, and tech companies invest in building the necessary infrastructure. As more devices become 5G-enabled, and industries continue to innovate, the potential applications of 5G will only expand, opening the door to new business models, services, and experiences.

Applications:
The 5G network is set to transform industries and everyday life. Here are some key applications:

- **Autonomous Vehicles:** 5G enables real-time communication between vehicles, traffic systems, and infrastructure, ensuring the safe and efficient operation of **self-driving cars**.

- **Virtual and Augmented Reality:** The increased speed and low latency of 5G allow for immersive **VR** and **AR** experiences in industries like **gaming**, **education**, and **healthcare**, enhancing user experiences without interruptions.

- **Industrial IoT:** Smart factories and logistics systems powered by 5G enable real-time communication between machines, optimizing production, reducing downtime, and enhancing efficiency across industries.

Summary:
Next-gen 5G networks are reshaping how we connect, work, and live by providing **ultra-fast speeds, low latency**, and **massive device connectivity**. From enabling autonomous vehicles and immersive virtual reality to creating smart cities and transforming healthcare, 5G offers a new era of connectivity that will revolutionize industries and daily life. Smart homes, telemedicine, and real-time data analysis are just a few of the areas where 5G is making a **profound impact**, improving efficiency, safety, and quality of life. As 5G continues to roll out, its influence will extend across the globe, powering the future of innovation and connectivity.

036 Smart Clothing
Technology Woven into Fabric

Imagine putting on a shirt that not only keeps you warm but can also **track your heart rate, monitor your posture**, and even **charge your smartphone**. Welcome to the world of **smart clothing**, where fashion meets cutting-edge technology. Over the past decade, smart clothing has evolved from concept to reality, revolutionizing the way we think about what we wear. Gone are the days when technology was limited to our wrists and pockets—now, it's woven directly into the fabric of our **clothes**.

The idea of smart clothing started with simple innovations, like fabrics that could change color or light up in response to external stimuli. But recent breakthroughs in **wearable technology** and e-textiles have taken this concept to a whole new level. By embedding **sensors, microchips**, and **conductive fibers** into fabric, researchers have developed clothing that can interact with the wearer's body and environment. These sensors can monitor a range of data, from **biometric information** like heart rate and body temperature to environmental conditions such as humidity and air quality.

One of the driving forces behind this revolution is the development of **conductive textiles**—fabrics that can transmit **electrical signals**. These materials allow designers to **integrate circuits, sensors**, and **power sources** directly into the clothing without affecting its comfort or appearance. For example, researchers have created shirts that can measure **electrical activity** in the heart, providing real-time data on the wearer's cardiovascular health. These garments offer the convenience of continuous health monitoring without the need for bulky external devices.

Beyond health applications, smart clothing is finding its way into the world of **sports** and **fitness**. Companies like Under Armour and Nike are developing **high-tech** athletic wear that can track **movement**, analyze **performance**, and provide feedback on **form** and **posture**. Imagine a pair of running shoes that not only records your distance and speed but also offers personalized coaching tips to improve your stride. Smart clothing is becoming an integral part of the fitness **tech revolution**, helping athletes of all levels push their limits.

One of the most exciting developments in smart clothing is the potential for **self-powering garments**. Researchers are working on fabrics that can **generate electricity** from the wearer's movement or body heat, effectively turning your clothes into a power source. This technology could be used to charge smartphones, power wearable devices, or even provide energy for heating or cooling the garment itself. The prospect of **self-sustaining, energy-harvesting** clothing opens up a world of possibilities for both consumer and industrial applications.

Applications:

Smart clothing has a wide range of applications that are already making an impact across several industries:

- **Healthcare:** One of the most promising applications of smart clothing is in **remote health monitoring**. Smart garments equipped with sensors can track vital signs such as heart rate, respiratory rate, and temperature, allowing for continuous health monitoring without the need for bulky medical devices. For patients with **chronic conditions** like cardiovascular disease or diabetes, smart clothing can provide doctors with **real-time data**, improving treatment outcomes and potentially saving lives.

- **Fitness:** Smart clothing is also transforming the fitness world. Wearable fitness trackers like smartwatches have become popular, but smart clothing takes this concept to the next level by offering more accurate data and **real-time feedback**. Compression shirts with **embedded sensors** can monitor muscle activity and fatigue, helping athletes prevent injuries. Smart yoga pants can track posture and alignment, guiding the user through their workout.

- **Military and Industrial Use:** In the military and industrial sectors, smart clothing offers solutions for **improving worker** and **soldier safety**. For instance, smart uniforms equipped with temperature-regulating fibers can adjust to extreme weather conditions, keeping soldiers comfortable in both hot and cold environments. Other garments can detect exposure to harmful chemicals or dangerous levels of fatigue, alerting the wearer and their team to potential hazards.

- **Fashion and Everyday Use:** Smart clothing isn't just for professionals or athletes—it's making its way into everyday fashion as well. Some garments are designed to **interact** with smartphones or smart home devices. For example, a smart jacket from Levi's integrates with Google's Project Jacquard, allowing users to control their phone's music or answer calls with a simple gesture. These innovations bring a new level of convenience to daily life.

Summary:
Smart clothing is reshaping the way we think about what we wear, blending fashion with **cutting-edge technology** to create garments that do more than just cover our bodies. From healthcare to fitness, smart clothing offers a range of **benefits**, including continuous health monitoring, enhanced athletic performance, and improved safety in industrial settings. As this technology evolves, the potential for self-powering garments that harvest energy from our movement or environment opens up even more possibilities. As the line between **technology** and **fashion** continues to blur, smart clothing is poised to become an integral part of our daily lives, making the future of fashion truly interactive.

037 3D Printing Homes
Revolutionizing Construction

The construction industry has remained largely **unchanged for decades**—until now. In the last decade, the **rise of 3D printing** has started to disrupt how we think about **building homes**, bringing a technological twist to a field known for bricks, mortar, and manual labor. Using **giant 3D printers**, houses are now being built **layer by layer**, offering a faster, cheaper, and more sustainable way to construct buildings. This innovative approach is not just an experimental idea but has already been put to the test, creating homes in a fraction of the time and at a fraction of the cost compared to traditional methods.

The concept is simple but groundbreaking. Just as a desktop 3D printer builds up layers of plastic to create objects, **large-scale 3D printers** can do the same with **concrete** or other **building materials**. A robotic arm deposits the material in precise **layers**, following a digital blueprint. This allows for extreme customization and flexibility in design. Complex curves, overhangs, and unique architectural elements that would be difficult or expensive to produce using traditional methods can now be created with **ease**.

One of the most exciting aspects of **3D printing homes** is its potential to address the global housing crisis. With a growing population and an increasing **demand** for affordable housing, traditional construction methods often fall short due to high costs and labor shortages. 3D printing offers a **solution** by drastically reducing construction **time** and **costs**. In some cases, 3D-printed homes can be built in less than **24 hours**. Companies like ICON and Apis Cor have already demonstrated the **viability** of this technology by constructing homes in under a day, at a cost that's significantly lower than traditional housing.

Beyond **speed** and **cost**, **sustainability** is another major advantage of **3D-printed homes**. Traditional construction generates massive amounts of waste, contributing to environmental degradation. By contrast, 3D printing is highly **efficient**, using only the amount of material needed to complete the structure. This reduces waste and lowers the environmental footprint of construction projects. Additionally, some companies are experimenting with using **recycled materials** or even **local soil** as building materials, further enhancing the sustainability of 3D-printed homes.

However, the technology is still in its **early stages**, and there are **challenges** to overcome. For one, **regulatory** hurdles remain, as building **codes** and **standards** often lag behind technological innovation. There's also the question of scalability—while a few homes can be printed quickly, can this technology be scaled to meet global demand? And while 3D printing dramatically reduces the need for **manual labor**, it also raises concerns about job displacement in the construction industry.

Applications:

The potential applications of 3D printing in the housing sector are vast, from providing affordable homes to improving disaster relief efforts:

- **Affordable Housing:** The most immediate and impactful application of 3D-printed homes is in addressing the **affordable** housing crisis. In places like Mexico and the U.S., 3D-printed houses are already being built for communities in need, offering **high-quality homes** at a fraction of the cost. With 3D printing, housing could become more **accessible** to people worldwide.

- **Disaster Relief:** 3D printing can play a crucial role in providing rapid housing **solutions** in the wake of **natural disasters**. Traditional methods often take months to rebuild homes after a disaster, but with 3D printing, shelters can be erected in a matter of **days**, providing immediate relief for displaced people.

- **Sustainability in Urban Development:** In rapidly growing cities, 3D-printed homes offer a sustainable alternative to traditional construction. By reducing **waste** and allowing for more **energy-efficient** designs, 3D-printed homes can contribute to greener urban development. Some projects are even exploring the use of **local materials** to print homes, further reducing the environmental impact.

Summary:
3D printing has the potential to **revolutionize** the construction industry by making homes **faster, cheaper**, and more **sustainable**. Companies around the world are already demonstrating the **viability** of **3D-printed homes**, with the technology showing promise in addressing the global housing crisis, providing rapid disaster relief, and improving urban sustainability. While **challenges** remain, particularly in scaling the technology and navigating regulatory issues, the future of 3D-printed homes is bright. As the technology continues to evolve, we may soon see a world where **homes** are **printed**, not built, bringing **new possibilities** to the construction industry.

038 Space Mining
Extracting Resources from Asteroids

In the past decade, the concept of **space mining** has transitioned from science fiction to a serious topic of discussion among scientists, governments, and private companies. With **Earth's natural resources** depleting and the demand for **metals, water,** and **minerals** increasing, the vast, untouched wealth floating in space—specifically in **asteroids**—has become an **attractive target** for future exploration and exploitation.

Asteroids, those rocky remnants left over from the formation of the solar system, are more than just space debris. Many of them are **rich** in valuable **materials** such as **iron, nickel, platinum**, and even **water**, which could be broken down into hydrogen and oxygen, providing fuel for space missions. Some estimates suggest that a single asteroid could contain more **precious metals** than have ever been mined on Earth, potentially worth trillions of dollars. This promise of untold wealth has sparked interest from both governments and private enterprises, hoping to tap into this **cosmic treasure** trove.

The idea of **mining asteroids** isn't just about bringing **precious metals** back to Earth. Space mining could play a crucial role in establishing a **sustainable** presence in **space**. Resources mined from asteroids could be used to build structures, fuel spacecraft, and support life on **future space colonies**. In essence, asteroids could be stepping stones to a self-sufficient space economy.

Several companies have already started planning missions to explore and eventually mine asteroids. **Planetary Resources**, one of the pioneers in the field, aims to survey asteroids and develop technology for extracting valuable resources. Another company, **Deep Space Industries**, envisions a future where asteroid mining supports not only **space exploration** but also **industry** and **commerce** in space.

NASA has also taken an interest in asteroid mining, not just for the **economic potential**, but for the scientific discoveries that could come from studying these ancient bodies. In 2016, NASA launched the **OSIRIS-REx mission**, which aims to collect samples from the **asteroid Bennu**. This mission is an important step toward understanding the composition of asteroids and how feasible mining them could be.

Yet, the **challenges** are significant. Mining an **asteroid** is a completely different challenge than mining on **Earth**. In space, **gravity** is weak or nonexistent, making it difficult to anchor equipment. Asteroids are often **far** from Earth, requiring long, **costly journeys** to reach them. And once there, the logistics of **extracting**, **processing**, and **transporting** materials back to Earth or other destinations are daunting.

Despite these obstacles, the potential **rewards** are enough to keep interest high. The global **demand** for rare Earth **elements**, **metals**, and other **materials** continues to rise, especially as industries like electronics, renewable energy, and electric vehicles grow. These sectors rely heavily on materials that are becoming more difficult and expensive to source on Earth. Space mining offers a potential **solution** to this resource crunch.

Applications:
While space mining may still be in its early stages, the possibilities it presents are vast and varied:

- **Construction in Space:** Rather than transporting building materials from Earth, **mined materials** from **asteroids** could be used to construct habitats, spacecraft, and other infrastructure in space. This would make long-term space missions, **lunar bases**, or even **Mars colonies** more feasible and cost-effective.

- **Fuel Production:** One of the most valuable resources on asteroids is **water**, which can be broken down into hydrogen and oxygen. These elements are critical for **rocket fuel**. Mining water from asteroids could **refuel spacecraft** while they are still in space, reducing the need to carry large amounts of fuel from Earth.

- **Economic Potential:** Asteroids contain **vast amounts** of precious metals like platinum, which are scarce on Earth. Mining these metals and bringing them back could **generate** trillions of dollars and **revolutionize** industries reliant on rare materials.

Summary:
Space mining offers a bold and ambitious **solution** to Earth's growing resource challenges. By extracting **valuable materials** such as metals and water from asteroids, we could support both Earth's economy and future space exploration. While the logistical and technological **challenges** are significant, the **potential rewards**—both scientific and economic—make it an exciting prospect. From **fueling** future space missions to building off-world **colonies**, space mining could open the door to a new era of human presence in **space**, with asteroids acting as the key to a self-sustaining space economy.

039 Human-Robot Collaboration
Merging the Strengths of Both

The future of work isn't about humans being replaced by robots, but about **collaboration** between the two. **Human-robot collaboration (HRC)** has gained traction in recent years, especially with the rise of **cobots**—short for **collaborative robots**. These robots are designed to work **safely** alongside **humans**, combining the precision and strength of machines with the creativity, problem-solving, and adaptability of humans. This partnership is already transforming various industries, from **manufacturing** to **healthcare** and agriculture.

One of the pioneers in this field is **Universal Robots**, a Danish company that developed some of the first cobots. Their **UR series**—like the **UR3**, **UR5**, and **UR10**—is used in countless factories worldwide. These cobots can perform repetitive, physically demanding tasks like assembly, painting, and welding, while human workers focus on supervising, troubleshooting, and ensuring quality control. The **robots** are equipped with **advanced sensors** that allow them to stop immediately if they detect a human in their path, ensuring a **safe working environment**. This innovation enables smaller manufacturers, who couldn't previously **afford industrial robots**, to adopt automation and remain competitive.

In automotive manufacturing, companies like BMW and Ford have embraced **human-robot collaboration** to streamline production. BMW's plant in Spartanburg, South Carolina, uses **KUKA** cobots to install insulation and doors in cars. By letting the cobots handle heavy lifting, BMW has reduced worker fatigue and injuries, while also speeding up production. Ford has been experimenting with collaborative **exoskeletons** for its employees, enabling them to lift heavy parts with ease. These devices, developed by companies like

Ekso Bionics, augment the worker's strength, reducing strain and injury, and making **human-robot collaboration** even more seamless.

Outside of manufacturing, human-robot collaboration is changing the **healthcare** landscape. Robots are assisting in **surgeries**, handling tasks that require **extreme precision**. **Intuitive Surgical's da Vinci robot**, for example, allows surgeons to perform minimally invasive surgeries with greater accuracy. Surgeons control the robot, which translates their hand movements into precise micro-movements, enabling them to work with a level of **precision** that's impossible with the human hand alone. This technology reduces recovery times for patients and minimizes scarring, revolutionizing procedures like prostatectomies and hysterectomies. Another exciting development is **Moxi**, a robot designed by **Diligent Robotics** that helps nurses with non-patient-facing tasks, like fetching supplies or delivering medication. This frees up healthcare professionals to focus more on direct patient care, making hospitals **more efficient**.

In **agriculture**, robots are making a significant impact by assisting with **labor-intensive tasks** like harvesting. **Iron Ox**, a California-based company, has developed autonomous farming systems where robots help plant, water, and harvest crops in highly controlled environments. Another example is **Agrobot**, which specializes in robots designed to pick **delicate fruits** like strawberries. These machines use cameras and sensors to identify ripe fruits and harvest them with the **gentleness** of a **human hand**, while still maintaining the speed and endurance that humans can't match. This approach helps farmers tackle labor shortages and improve productivity.

Despite the progress, integrating robots into the workplace comes with **challenges**. One of the primary concerns is **trust**—can workers trust robots to operate safely alongside them? This is why companies invest heavily in **sensors** and **AI systems** that ensure robots stop immediately if a human enters their work zone. **Rethink Robotics**, before being acquired by Hahn Group, developed the famous cobot **Baxter**, which featured a highly intuitive interface and built-in safety systems. Baxter was designed to interact with human workers, adapting to their needs and learning on the job through demonstration, rather than being programmed.

Additionally, workers must be retrained to **collaborate** with **robots**. Learning how to program, operate, and troubleshoot robots has become a necessary skill in many industries. Far from replacing jobs, **robots** are creating **new**

roles that require humans to engage in higher-level tasks, while robots handle the repetitive and physically demanding aspects. Companies like **FANUC** and **ABB Robotics** are leading the charge by offering training programs for workers to develop these skills, ensuring the workforce is prepared for this new **collaborative future**.

Applications:
Human-robot collaboration is already being implemented in several key industries:

- **Manufacturing:** At BMW's Spartanburg plant, **cobots** from **KUKA** install car doors and insulation, taking over repetitive, heavy lifting tasks. Meanwhile, workers ensure **quality control** and **problem-solving**. This combination has increased efficiency and reduced injuries on the assembly line.

- **Healthcare:** The **da Vinci robot by Intuitive Surgical** enables surgeons to perform minimally invasive surgeries with unmatched **precision**. Surgeons control the robot, allowing for highly accurate procedures, reducing recovery times, and improving patient outcomes. Meanwhile, robots like **Moxi** assist nurses with routine tasks, letting them focus more on patient care.

- **Agriculture:** Companies like **Iron Ox** and **Agrobot** are using robots to assist in **planting**, **watering**, and **harvesting** crops. Agrobot's fruit-picking robots carefully harvest delicate produce like strawberries, increasing yield and reducing labor shortages.

Summary:
The era of **human-robot collaboration** is here, and it's transforming industries across the board. From manufacturing plants using **KUKA cobots** to improve efficiency, to hospitals deploying robots like **Moxi** and the **da Vinci system** to assist in surgeries and patient care, **humans** and **robots** are working together more than ever before. The goal isn't to replace workers, but to augment their **capabilities**—robots take care of repetitive, physically demanding tasks, while humans focus on **creativity**, **problem-solving**, and critical **decision-making**. This harmonious collaboration enhances productivity, worker safety, and innovation, signaling a future where the strengths of humans and robots are combined to shape a new, **more efficient world**.

040 Digital Twins
Simulating the Real World in Digital Form

In the last decade, one of the most exciting and transformative innovations in technology has been the rise of **digital twins**. These are not just advanced simulations; they are exact, **real-time** digital **replicas** of physical objects, systems, or processes. Imagine having a **digital counterpart** of a complex system—whether it's a factory, a wind turbine, or even a human body—that you can **interact** with and monitor, making **adjustments** in real time **without affecting** the actual physical entity. That's the **power** of **digital twins**.

The **concept** was first introduced in the early **2000s**, but it's only recently, with **advancements** in **cloud computing, big data,** and **AI**, that digital twins have become practical and widely adopted. Digital twins allow us to bridge the gap between the **physical** and **digital** worlds by creating a **virtual representation** of a physical object. **Sensors** attached to the real-world object **feed data** into the digital twin, allowing it to **replicate** and **predict** behavior in real-time. This is done with stunning **accuracy**, meaning you can run simulations, test scenarios, and even **predict** future issues, all within the virtual space.

Initially used in high-tech industries like aerospace and manufacturing, **digital twin** technology has rapidly **expanded** into other **sectors**.
For example, companies like General Electric (GE) and Siemens have adopted **digital twins** to **monitor** their **industrial equipment,** while the automotive and healthcare sectors are using digital twins to improve everything from car design to personalized medicine. The goal is to **replicate** the **physical world** in enough **detail** that users can **test, troubleshoot,** and **optimize** systems in real time.

One of the **greatest advantages** of digital twins is their ability to **anticipate problems** before they happen. Take a jet engine, for instance. With a digital twin, engineers can monitor every aspect of the engine's performance while it's still in use. They can simulate how it will react under various conditions, make **real-time adjustments**, and even predict when maintenance will be required. This predictive maintenance can **save** companies **millions** by preventing equipment failure and reducing downtime.

Another exciting application is in **smart cities**, where digital twins are used to **monitor** and **manage** urban infrastructure. Entire cities can be replicated in digital form, allowing city planners and governments to **optimize** traffic flow, manage energy consumption, and even **simulate** responses to natural disasters. These digital replicas can help cities become more sustainable, efficient, and responsive to the needs of their citizens.

Applications:

The versatility of digital twins allows them to be applied across numerous industries, offering real-time monitoring, predictive maintenance, and enhanced decision-making. Here are some of the most impactful ways digital twin technology is being used today:

- **Manufacturing:** In the industrial world, digital twins are used to monitor machinery and equipment in real-time. By running simulations on a virtual model, engineers can predict failures, optimize performance, and schedule maintenance before a machine breaks down, saving both time and money.

- **Healthcare:** In the medical field, digital twins of human organs or entire body systems can be created. This allows doctors to simulate different treatment plans on a digital replica of a patient's body before administering the treatment in real life, providing a personalized healthcare solution. Imagine testing how different medications might affect a heart condition using a digital twin of the patient's heart.

- **Urban Planning:** Smart cities are increasingly turning to digital twin technology to model and manage their infrastructure. By creating digital replicas of roads, buildings, and public transportation systems, city planners can run simulations to improve traffic flow, reduce energy use, and prepare for natural disasters like floods or earthquakes.

- **Automotive Design:** The automotive industry uses digital twins to simulate vehicle designs and test them under various conditions. By running virtual crash tests or checking how a vehicle will perform in extreme weather conditions, manufacturers can make better, safer cars before a physical prototype is even built.

Summary:
Digital twins are revolutionizing industries by providing a **real-time, virtual** representation of **physical** objects and systems. From jet engines to smart cities, the ability to **monitor, simulate,** and **predict** outcomes in a **digital environment** is opening new frontiers in efficiency, sustainability, and personalized solutions. The widespread adoption of digital twin technology is already helping industries **save** billions in **maintenance** and **operational costs**, while also paving the way for smarter, more efficient urban planning and healthcare. The future of **digital twins** is bright, as they continue to blur the lines between the **physical** and **digital worlds**, offering a seamless integration that could change how we interact with **technology** on a day-to-day basis.

041 Synthetic Biology
Designing Life in the Lab

Over the last decade, the boundaries of **biology** have been pushed in ways that would have once seemed like **science fiction**. At the forefront of this **revolution** is **synthetic biology**, a field that merges biology with engineering to design and construct new biological **parts**, systems, and **organisms**. Unlike traditional genetic engineering, which typically modifies existing organisms, synthetic biology seeks to **create** entirely **new forms** of life from scratch, offering the possibility to reprogram living systems with unprecedented **precision**.

The concept of synthetic biology is rooted in the idea that **life** can be **engineered** in much the same way that **machines** are **built**. Scientists have begun to apply engineering principles—**standardization**, **modularity**, and **automation**—to biology. In doing so, they have developed **biological circuits** that can be inserted into **living cells** to control their behavior, just as computer circuits control a machine. By designing and constructing **DNA sequences** that act as "**programming code**", researchers can rewire cells to perform new tasks—whether that's producing biofuels, cleaning up pollutants, or even detecting diseases.

One of the most significant breakthroughs in synthetic biology came with the creation of **synthetic genomes**. In 2010, researchers at the **J. Craig Venter Institute** successfully created the first self-replicating, **synthetic bacterial cell**. This organism, called **Mycoplasma laboratorium**, was built entirely from **artificially constructed DNA**. While it resembled natural life, every piece of its genetic code was designed and assembled in the **lab**, marking a major milestone in humanity's ability to design **life**.

Since then, the field has rapidly advanced. Scientists are now working on creating **synthetic organisms** that can be used to manufacture **pharmaceuticals, materials,** and **chemicals** in a more sustainable and efficient way. These bioengineered organisms hold the promise of reducing our reliance on petrochemicals and other environmentally harmful resources. Additionally, synthetic biology has opened up new avenues for **medical therapies**, with researchers working to create **custom-designed** organisms that can seek out and destroy **cancer cells** or deliver **targeted drugs** within the human body.

However, like many groundbreaking fields, synthetic biology comes with **ethical challenges** and **concerns**. The ability to design and create new forms of life raises questions about the **potential risks**, such as unintended consequences if synthetic organisms were to **interact** with **natural ecosystems** in unpredictable ways. There are also concerns about **bioterrorism**, where synthetic organisms could be weaponized for harmful purposes. As a result, there has been increasing focus on establishing **ethical guidelines** and **regulatory frameworks** to ensure that this powerful technology is used responsibly.

Applications:

The potential applications of synthetic biology span across various industries, and many are already making a significant impact:

- **Biomanufacturing:** Synthetic biology allows for the creation of **microbes** that can produce **valuable materials** like biofuels, bioplastics, and pharmaceuticals. These synthetic organisms are designed to be more **efficient** and **environmentally friendly** than traditional manufacturing processes. For example, synthetic yeast is being engineered to produce **artemisinin**, an important drug for treating **malaria**, at a much lower cost.

- **Environmental Remediation:** Synthetic organisms are being designed to tackle some of the most pressing **environmental challenges**. Scientists have developed bacteria that can break down plastic waste in oceans or clean up oil spills, offering new tools for environmental protection and sustainability.

- **Medicine:** In the medical field, synthetic biology is being used to develop **new therapies** for diseases like **cancer**. Custom-built bacteria are being engineered to target and kill **tumor cells** while leaving healthy tissue unharmed. These living medicines can be programmed to respond to specific signals in the body, providing a level of **precision** that conventional treatments cannot match.

Summary:
Synthetic biology is redefining what it means to "create life". By designing biological systems in the lab, scientists are unlocking **new possibilities** in everything from manufacturing to medicine. With applications ranging from biofuels to cancer treatments, synthetic biology holds the promise of addressing some of the world's **most critical challenges**. However, with this great power comes the responsibility to navigate the **ethical** and **environmental** risks associated with creating new life forms. As the field continues to evolve, synthetic biology stands as one of the most exciting and transformative developments of the **last decade**.

042 Discovery of Quantum Entanglement Applications
Instant Connections

In the early 20th century, Albert Einstein famously referred to **quantum entanglement** as "spooky action at a distance". At the time, even the brightest minds in physics struggled to fully comprehend how **two particles**, separated by vast distances, could be **instantaneously connected**. But over the last decade, we've moved from theoretical musings to real-world applications of this mind-bending phenomenon.

Quantum entanglement occurs when two particles become **interconnected** in such a way that the **state** of one **particle** directly affects the state of the other, no matter how **far** apart they are. This means that if you were to measure one particle, you would instantly know the state of its entangled partner, even if it were on the other side of the universe. The **speed** of this **interaction** seemingly violates the fundamental cosmic **speed limit**: the speed of **light**. However, no actual information is traveling faster than light; it's the particles' shared **quantum state** that creates this **instantaneous connection**.

For years, quantum entanglement remained a fascinating but largely **impractical idea**. However, in the past decade, scientists have begun discovering tangible **applications** for this seemingly **magical** effect, especially in the field of **quantum communication**. Imagine a future where you could send encrypted messages instantly across **vast** distances, entirely **secure** and **immune** to hacking. This is the promise of quantum teleportation and **quantum key distribution (QKD)**, technologies that could revolutionize the way we communicate, ensuring absolute **privacy** and **security**.

One of the most groundbreaking advancements in this area came in 2017 when Chinese researchers successfully demonstrated **quantum teleportation** between Earth and a satellite orbiting 1,200 kilometers above. Using quantum entanglement, they were able to transfer the **quantum state** of a photon on Earth to a photon in space without any physical connection between them. This experiment marked a significant milestone, showing that **quantum communication** on a global scale could one day become a **reality**.

The key to this futuristic form of **communication** lies in the fact that any attempt to intercept or tamper with the quantum-entangled particles would immediately **break** the **entanglement**, alerting the parties involved to the breach. This means that **quantum encryption** offers a level of security that is fundamentally unbreakable by classical computing methods. It's not just secure—it's **quantum secure**.

Applications:
While quantum entanglement may sound like something straight out of science fiction, its practical applications are already taking shape. Here are some of the most promising uses for this groundbreaking phenomenon:

- **Quantum Communication:** The idea of transmitting information across **long distances** without delay is becoming more feasible with the development of **quantum networks**. These networks could enable **secure** communication between distant locations, immune to the eavesdropping risks that plague classical networks today. For instance, countries like China are leading the charge by building quantum communication **satellites** that could one day enable **unhackable** global internet systems.

- **Quantum Cryptography:** Traditional cryptographic methods rely on **mathematical algorithms** to secure data, but these systems are vulnerable to quantum computers, which can **break** many of today's **encryption** techniques. With **quantum key distribution (QKD)**, encryption becomes impenetrable because any attempt to intercept the quantum key would **disrupt** the **entanglement**, alerting the communicating parties and making the interception impossible.

- **Quantum Teleportation:** While we may not be "**beaming**" people across the **universe** anytime soon, quantum teleportation allows for the **instantaneous transfer** of quantum states between particles. This has massive implications for **quantum computing**, where entanglement can be used to **synchronize qubits** across vast distances, enabling super-fast computation and communication.

Summary:
The discovery and application of **quantum entanglement** in the last decade have brought us closer to a world where **instant, secure** communication is not only possible but fundamentally **unbreakable**. From **quantum communication** to **quantum cryptography**, entanglement is poised to revolutionize the way we exchange information. The practical uses of quantum teleportation may be in their infancy, but the promise of a future where data transfer is **instantaneous** and **unhackable** is rapidly becoming a **reality**. This "spooky action" is no longer confined to theoretical physics; it's actively shaping the next generation of technology, offering new possibilities that challenge our understanding of **reality**.

043 Advances in Space Telescopes
Peering Deeper into the Cosmos

The last decade has seen remarkable **advancements** in **space telescopes**, revolutionizing how we explore the **universe**. These technological leaps have allowed us to **peer deeper** into space than ever before, revealing new insights about the cosmos that were previously beyond our reach. From unraveling the mysteries of distant **galaxies** to detecting **exoplanets** in other solar systems, modern space telescopes are redefining our understanding of the **universe**.

One of the most significant developments in recent years is the upcoming launch of the **James Webb Space Telescope (JWST)**. Set to be the **most powerful** space telescope ever created, the JWST will succeed the **Hubble Space Telescope** and is designed to see even further into the **universe's past**. While Hubble revolutionized space observation by capturing awe-inspiring images of **stars**, **nebulae**, and **galaxies**, JWST will take things even further. Unlike Hubble, which primarily observes in visible and ultraviolet light, the JWST is optimized to observe in the infrared spectrum, allowing it to see through **cosmic dust** and capture clearer images of **distant objects** that are otherwise obscured.

The JWST is designed to answer some of the most profound questions in astronomy: How did the **first galaxies** form after the **Big Bang**? Are there potentially **habitable planets** in other solar systems? How do **stars** and **planetary systems** evolve over time? To answer these questions, the telescope features a 6.5-meter **primary mirror**—much larger than Hubble's 2.4-meter mirror—giving it unparalleled sensitivity and resolution. Its instruments will allow it to detect faint light from the **earliest stars** and

galaxies, effectively looking back in time to when the universe was just a few hundred million years old.

Another groundbreaking space telescope is **TESS (Transiting Exoplanet Survey Satellite)**, launched in 2018. TESS has a specific mission: to search for **exoplanets**—planets outside our solar system—by monitoring the **brightness** of **stars**. When a planet passes in front of its star, known as a transit, it causes a slight dip in the star's brightness, which TESS can detect. Over the past few years, TESS has discovered **thousands** of potential **exoplanets**, some of which are located in the **habitable zone**, where **liquid water** could exist, raising the possibility of **life** beyond Earth.

TESS builds on the legacy of **Kepler**, the earlier exoplanet-hunting telescope, but is far more advanced. While Kepler focused on a small section of the sky, TESS surveys nearly the **entire sky**, providing a more comprehensive view of our **galactic neighborhood**. The discoveries made by TESS have captivated both scientists and the public, fueling interest in the search for extraterrestrial life.

The advancement of space telescopes is not just about seeing farther; it's about seeing more clearly and gathering more **data**. With the combination of new instruments, such as **spectrometers** and advanced **imaging systems**, we can now analyze the **atmospheres** of distant **exoplanets**, detect **black holes**, and explore phenomena like **dark matter** and **dark energy**—forces that make up the majority of the universe but remain poorly understood.

Applications:
The cutting-edge technology behind modern space telescopes like the JWST and TESS has practical implications beyond pure exploration:

- **Understanding the Origins of the Universe:** By peering deep into space, the James Webb Space Telescope will help astronomers study the formation of the **first galaxies**, shedding light on the origins of the universe. Observing these **distant galaxies** helps us understand how the universe evolved from the **Big Bang** to its current state.

- **Exoplanet Discovery and Habitability:** TESS and other space telescopes are revolutionizing the search for **exoplanets**. By detecting planets in the **habitable zones** of stars, scientists can focus on finding worlds that could potentially support **life**. This has led to new missions aimed at exploring these planets in greater detail.

- **Astrobiology and the Search for Life:** The discoveries made by these telescopes drive the field of astrobiology, the study of **life beyond Earth**. As we detect planets that could host life, the next step is to analyze their atmospheres and surfaces for signs of biological activity, such as water, oxygen, or methane.

- **Technological Innovations:** The engineering breakthroughs needed to build telescopes like JWST have broader applications. The same technologies used for precision **imaging** and **data** processing in space can be adapted for use on **Earth**, from improving medical imaging devices to enhancing communication systems.

Summary:

Advances in space telescopes over the last decade have dramatically changed our view of the **universe**. The **James Webb Space Telescope** promises to unlock new knowledge about the **early universe**, while **TESS** has revolutionized the hunt for **exoplanets**, bringing us closer to answering the question of whether **life** exists beyond Earth. These space telescopes not only help us look back in time and across vast distances but also enable practical advancements in fields like **astrobiology, cosmology**, and **technology**. As we continue to push the boundaries of what telescopes can do, the future of space exploration looks more promising—and awe-inspiring—than ever.

044 Solar Probe Missions
Touching the Sun

In 2018, NASA launched the **Parker Solar Probe**, a mission unlike any other. Its goal? To "touch" the **Sun**—something humanity had never done before. Over the next several years, Parker would travel **closer to the Sun** than any previous spacecraft, plunging into the outermost **layers** of its scorching atmosphere, the **corona**, to gather unprecedented data. At its closest approach, the probe is flying at a mere 4 million miles from the Sun's surface, a distance so near that temperatures around the spacecraft reach nearly **2,500 degrees** Fahrenheit. Yet, thanks to innovative **heat-shield** technology, Parker can withstand this intense environment, sending back invaluable data to Earth.

Why would we want to send a spacecraft so **dangerously** close to the Sun? Understanding the **Sun's behavior** is crucial because it directly affects **life on Earth**. The **solar wind**, a stream of charged particles emitted by the Sun, can interfere with **satellites**, **GPS**, and even **power grids**. More extreme solar events, like **coronal mass ejections (CMEs)**, can cause **geomagnetic storms** capable of disrupting entire power networks. By studying the solar wind and the corona up close, the Parker Solar Probe is helping scientists unlock the **mysteries** of these **phenomena** and improve space weather forecasting.

But the mission isn't just about protecting technology. By getting up close and personal with the **Sun**, the probe is also helping scientists answer fundamental questions about the **solar system**. One of the greatest mysteries is why the **Sun's corona** is so much **hotter** than its surface—sometimes by millions of degrees. This paradox, known as the **coronal heating problem**, has baffled scientists for decades.

Another key question the Parker probe is tackling is how the **solar wind** accelerates to such **high speeds** as it moves away from the Sun. Understanding these processes could give us deeper insight into not only our Sun but also **stars** throughout the **universe**.

What makes the Parker Solar Probe particularly exciting is the **scale** and **ambition** of the mission. It's traveling faster than any man-made object in history, reaching speeds of up to **430,000 miles per hour**—fast enough to travel from New York to Tokyo in under a minute. Every orbit brings it closer to the Sun, with a total of **24 passes** planned over seven years. As the spacecraft inches closer, it will brave ever more **extreme conditions**, all the while sending back data that could reshape our understanding of the **Sun** and its role in the **solar system**.

Applications:
The knowledge gained from the Parker Solar Probe has profound implications across multiple fields:

- **Space Weather Prediction:** The data collected by the probe is helping scientists improve **predictions** of **space weather**. Better forecasting of solar storms could help **protect** our technological **infrastructure** on Earth, such as power grids and communication satellites, and even safeguard **astronauts** in space from dangerous radiation.

- **Solar Power:** Understanding the **Sun's energy** output more precisely could have long-term benefits for solar power generation. As we shift towards **renewable energy**, solar power is a key player, and more knowledge about the Sun's activity will help **optimize** energy capture methods.

- **Advancing Astronomy:** The Parker Solar Probe's findings are not limited to our Sun. By learning about the Sun's processes, we gain insights that can be applied to other **stars**. This could deepen our understanding of **star formation**, **solar winds** in other **star systems**, and **stellar evolution**.

Summary:
The Parker Solar Probe mission is a **historic leap** in space exploration, providing humanity's first close-up look at the **Sun**. By venturing into the **corona**, the probe is solving long-standing mysteries, such as the **coronal heating** problem and the acceleration of the **solar wind**. The mission's data is already revolutionizing **space weather prediction** and could have applications

in fields as diverse as renewable energy and astronomy. As Parker continues its daring journey, it brings us one step closer to fully understanding the **star** that sustains life on **Earth**—our Sun.

045 Dark Matter and Dark Energy Studies
Unveiling the Invisible Universe

Over the past decade, some of the most profound discoveries in physics have revolved around what we cannot see. **Dark matter** and **dark energy** make up roughly **95% of the universe**, yet they remain some of the greatest mysteries in modern science. Despite their **invisible nature**, their effects are undeniable—they govern the structure, formation, and expansion of the universe. But what exactly are **dark matter** and **dark energy**, and why are scientists so focused on understanding them?

Dark matter is a form of matter that doesn't **emit, absorb,** or **reflect** light, making it **invisible** to current detection methods. Scientists know it exists because of its **gravitational effects**. Galaxies, for example, rotate at speeds that would tear them apart if only **visible matter** were present. Something else, something unseen, is providing the additional **gravitational pull** to hold these galaxies together. This **"something"** is what we call **dark matter**. The hunt for direct evidence of **dark matter** has been a driving force behind many experiments, from underground detectors to space-based telescopes, but it continues to elude direct observation.

On the other hand, **dark energy** is an even more **enigmatic** force. In the 1990s, astronomers made a shocking discovery: the **expansion of the universe** is accelerating. This was completely unexpected—gravity, which **pulls objects** together, should have been slowing the expansion down. To account for this mysterious acceleration, scientists introduced the concept of **dark energy**, a force that pushes the universe apart. While **dark matter** holds galaxies together, **dark energy** is driving the **universe** to expand faster and faster.

The **Lambda Cold Dark Matter (ΛCDM)** model, which is the most widely accepted cosmological **model**, incorporates both dark matter and dark energy to **explain** the **structure** and **evolution** of the universe. According to this model, only about 5% of the universe is made of ordinary, **visible** matter (like stars, planets, and galaxies), while **26%** is **dark matter** and **69%** is **dark energy**. However, despite knowing the proportions, the exact nature of dark matter and dark energy remains one of the biggest **unsolved mysteries** in physics.

Over the last decade, significant strides have been made in our understanding of these phenomena. The **Large Hadron Collider (LHC)**, for example, has been searching for potential dark matter particles by recreating conditions similar to those just after the **Big Bang**. Similarly, telescopes like the **Dark Energy Survey and Hubble Space Telescope** are constantly observing the **cosmos** to track the universe's expansion, providing clues to the **nature** of **dark energy**. However, as of now, scientists have yet to directly observe dark matter particles or fully explain the nature of dark energy, but the race is on.

Applications:
While dark matter and dark energy studies are more theoretical, they have far-reaching implications for our understanding of the universe:

- **Astrophysics:** The study of **dark matter** helps scientists explain how galaxies form and evolve. Understanding dark matter's role in

    cosmic structures could lead to new insights into the universe's origins and ultimate fate.

- **Space Exploration: Dark energy** research is essential for understanding the **universe's expansion**. If the universe continues to expand at an accelerating rate, it could affect future space missions and long-term exploration strategies.

- **Fundamental Physics:** The search for **dark matter particles** might lead to the discovery of **new particles** and **forces**, fundamentally altering our understanding of the laws of physics. This could revolutionize fields like **particle physics** and **cosmology**.

Summary:
Dark matter and dark energy are the unseen forces shaping the universe, making up 95% of everything in existence. While dark matter holds galaxies together, dark energy is pushing the universe apart, causing its expansion to accelerate. Over the past decade, significant progress has been made in studying these mysterious forces, although their true nature remains elusive. Experiments like those conducted at the Large Hadron Collider and observations from the Dark Energy Survey are providing new insights, but the universe still holds many secrets. As scientists continue to search for answers, unlocking the mysteries of dark matter and dark energy could reshape our understanding of the cosmos and the very nature of existence.

046 Higgs Boson Found
The Missing Piece of the Particle Puzzle

In July 2012, the world of physics celebrated a monumental achievement: the discovery of the **Higgs boson** at **CERN's Large Hadron Collider (LHC)**. This **elusive particle** had been theorized nearly 50 years earlier by physicist **Peter Higgs** and others as part of the **Standard Model of particle physics**, which explains how the basic building blocks of the universe interact. But until its discovery, the Higgs boson was the **missing piece** that completed the puzzle of the universe's fundamental forces.

The Higgs boson is **crucial** because it is associated with the **Higgs field**, an invisible field that permeates the entire universe. When particles **interact** with this field, they **gain mass**. Without the Higgs field, particles would zip around at the speed of light, unable to clump together to form atoms, stars, or planets. In essence, the Higgs boson is **responsible** for giving **mass** to matter, making the universe as we know it possible.

Detecting the Higgs boson **required** enormous **energy**, which is why it took until 2012, when the **Large Hadron Collider**—the world's most powerful particle accelerator—became operational. The LHC **smashes protons** together at nearly the **speed of light**, recreating conditions similar to those just after the **Big Bang**. When these collisions occur, they produce massive amounts of **energy**, sometimes creating new particles, like the Higgs boson, which can then be detected by the **sophisticated sensors** lining the collider's tunnels.

This discovery didn't just confirm decades of theoretical work; it also **validated** the **Standard Model**—the framework that describes how particles like **electrons**, **quarks**, and **photons** interact through **fundamental forces**. ithout the Higgs boson, the Standard Model would have been **incomplete**, and our understanding of the universe would have remained fundamentally flawed.

However, while the Higgs discovery filled a **critical gap** in the Standard Model, it also opened new questions about the universe. The Standard Model, despite its success, doesn't explain some of the most profound **mysteries**, such as the nature of **dark matter** or **dark energy**, which make up most of the universe's mass and energy. Physicists hope that continuing research at CERN and other facilities will shed light on these enigmas.

Applications:
While the discovery of the Higgs boson might seem far removed from everyday life, it has profound implications for our understanding of the universe and the technology that could emerge from it. Here are some ways this breakthrough is impacting science and technology:

- **Fundamental Physics:** The discovery of the Higgs boson provides critical **confirmation** of the **Standard Model**, which underpins much of modern physics. With this knowledge, physicists can now focus on probing deeper **mysteries** like **dark matter** and the unification of all fundamental forces.

- **Particle Accelerators and Technology:** The technology developed to find the Higgs boson—particularly the **detectors** and **particle accelerators**—could have applications beyond physics. For example, medical imaging technologies like **PET** scans already use particle physics principles, and advances in particle **detectors** could lead to improvements in healthcare diagnostics.

- **Quantum Field Theory:** The discovery of the Higgs boson has deepened our understanding of **quantum fields**, which describe how **forces** and **particles** interact on a fundamental level. This knowledge could pave the way for new technologies based on **quantum mechanics**, such as **quantum computing** or advancements in materials science.

Summary:
The discovery of the **Higgs boson** in 2012 was a monumental moment in the history of science, completing the **Standard Model** and confirming the existence of the **Higgs field**, which gives **mass** to particles. While this discovery answered a fundamental question about the **universe's structure**, it also opened the door to new mysteries about **dark matter**, **dark energy**, and the deeper workings of the

cosmos. The technology developed to detect the Higgs boson is already influencing fields like medical imaging, and as we continue to explore the **universe** at its most fundamental level, the potential for new technologies is boundless.

047 Discovery of Liquid Water on Mars
Possibility of Life

In 2018, NASA made an announcement that reignited humanity's curiosity about **Mars**: they had discovered evidence of **liquid water** beneath the **planet's surface**. Using radar data from the **European Space Agency's Mars Express** orbiter, scientists detected a **subglacial lake** more than a mile beneath the planet's south polar ice cap. This wasn't just ice or vapor—it was **liquid water**, a key ingredient for **life** as we know it.

For decades, scientists had speculated about the presence of **water** on **Mars**. In fact, Mars' surface shows signs that **water** once **flowed** there—**dry** **riverbeds, sediment layers**, and **ancient deltas** are scattered across the planet. But discovering **liquid water** in the present day was a game-changer. It suggested that, even in the harsh conditions of **Mars**, some form of water-based activity could be taking place right now, which opens up exciting possibilities about life on the **Red Planet**.

Mars is a **cold** and barren world, with average **surface temperatures** well below freezing. The **thin atmosphere**, composed mostly of carbon dioxide, offers little **protection** from **radiation** or the extreme **cold**. This is why the **discovery** of **water** beneath the surface was so surprising—liquid water shouldn't be able to exist under such conditions.

However, it's believed that **salts** mixed into the water lower its freezing point, allowing it to **remain liquid** despite the frigid temperatures. The **lake** is estimated to be about **12 miles wide**, but scientists don't know how deep it is or what lies beneath it. What they do know is that wherever there's liquid water, there's a chance for **life**. On **Earth, microbial life** has been found thriving in some of the most extreme environments—deep below glaciers in **Antarctica**, in the high-pressure depths of the ocean, and in volcanic hot springs. If life can exist in these extreme conditions on **Earth**, could it also exist on **Mars**?

The discovery of **liquid water** on Mars doesn't guarantee that life exists there, but it certainly increases the chances. For now, the next steps involve further exploration and, potentially, sending **missions** to drill beneath the Martian surface to study the water directly. This could provide answers about the lake's composition, its salinity, and, most importantly, whether there are any signs of **microbial life** lurking beneath Mars' frozen crust.

Applications:

The discovery of liquid water on Mars has significant implications for both scientific exploration and the future of human space travel:

- **Search for Life:** The presence of **liquid water** beneath the Martian surface gives scientists a promising place to look for **microbial life**. Just as life on Earth thrives in extreme environments, such as deep under glaciers, it's possible that similar life forms could exist on **Mars**. Future missions could focus on drilling into the Martian ice to analyze the water for **organic molecules** or **microbial signatures**.

- **Human Colonization:** If humans are ever going to **colonize** Mars, access to **water** is essential. Water can be used for **drinking, growing food**, and even producing **fuel** through **electrolysis**, a process that separates water into oxygen and hydrogen. The discovery of liquid water means that Mars could be more hospitable to future **colonies** than previously thought.

- **Further Exploration:** This discovery provides a clear target for future **Mars missions**. NASA, along with other space agencies, can now prioritize exploring the Martian poles, where **liquid water** seems to exist. By studying Mars' water in greater detail, we can learn more about the planet's **geological history** and its potential to support **life**.

Summary:
The discovery of **liquid water** on **Mars** in 2018 has reignited the possibility that **life** might exist beyond Earth. While the planet's harsh conditions make life unlikely on the surface, the presence of a subglacial lake beneath Mars' south pole offers **hope**. **Water** is a **fundamental** ingredient for **life**, and where water exists, so too might life, even in the form of resilient **microbes**. This discovery not only expands our understanding of Mars' potential **habitability** but also has huge implications for future exploration and **human colonization**. As missions to Mars continue, the dream of uncovering life on the **Red Planet** inches closer to reality.

048 Breakthrough Listen Project
Searching for Extraterrestrial Life

In 2015, a monumental initiative was launched to tackle one of humanity's most enduring questions: **Are we alone in the universe?** The **Breakthrough Listen Project**, spearheaded by physicist **Stephen Hawking** and Russian entrepreneur **Yuri Milner**, is the most comprehensive search for extraterrestrial life ever undertaken. With an unprecedented **budget** of $100 million over 10 years, this initiative aims to scan the skies for **signals** from **intelligent civilizations** beyond Earth.

The scope of the Breakthrough Listen Project is staggering. It focuses on the search for **technosignatures**, or signs of advanced technologies, from planets orbiting distant stars. Using the world's most powerful **telescopes**, including the **Green Bank Telescope** in West Virginia and the **Parkes Telescope** in Australia, the project scans millions of stars and over 100 nearby galaxies. The project covers a much broader range of **radio frequencies** than previous attempts to **detect alien civilizations**, which were limited in scope due to technology and funding constraints.

What sets Breakthrough Listen apart from previous efforts, such as the **SETI (Search for Extraterrestrial Intelligence)** program, is not only its scale but also the use of cutting-edge technologies. The project leverages **machine learning** and **AI** to sift through the vast amounts of **data collected** by these massive telescopes. With this technology, Breakthrough Listen can search for **patterns** or **anomalies** in the **radio signals** that might indicate the presence of **intelligent life**, even across millions of data points.

The project also invites collaboration from the public and other researchers. Through platforms like **SETI@home**, people can volunteer their computers' processing power to help analyze the enormous amount of data. This level of community involvement, coupled with the most advanced **data analysis techniques**, makes Breakthrough Listen the most comprehensive and inclusive search for extraterrestrial life ever conducted.

One of the most exciting aspects of the project is that it doesn't focus solely on radio signals. Breakthrough Listen also **scans** the skies for **optical signals**, such as laser pulses, which could be used by extraterrestrial civilizations as a form of **communication**. By covering both the radio and optical spectrum, the project ensures that it isn't missing any **potential signals**.

Though no definitive signs of extraterrestrial life have been found yet, the **Breakthrough Listen Project** has already uncovered interesting **signals** that require further investigation. These signals, known as narrowband radio bursts, could potentially be technosignatures, but researchers must first rule out **natural phenomena** or **interference** from human-made objects like satellites.

Applications:

The search for extraterrestrial life is not just about satisfying human curiosity—it has profound implications for our understanding of the universe and our place in it. Here are a few key areas where the Breakthrough Listen Project could make a significant impact:

- **Advances in Technology:** The **technological innovations** developed for this project, such as **machine learning algorithms** for processing vast amounts of data, can have applications beyond the search for aliens. These technologies are being adapted for use in fields like astronomy, climate modeling, and healthcare to analyze complex data sets.

- **Understanding Cosmic Phenomena:** Even if we don't find **alien life**, the data collected by Breakthrough Listen could lead to new **discoveries** about the **universe**. By scanning the skies so extensively, scientists may uncover previously **unknown cosmic phenomena**, such as new types of stars, pulsars, or black holes.

- **International Collaboration:** The global nature of this project has fostered unprecedented **collaboration** between countries and organizations. This cooperative effort has improved **global scientific diplomacy**, showing that the search for life beyond Earth can bring humanity together, even across borders.

Summary:
The **Breakthrough Listen Project** is humanity's most ambitious **search** for **extraterrestrial life**, using the latest technology to **scan** the **universe** for signs of intelligent civilizations. With a vast budget, access to the most powerful **telescopes**, and the use of **AI** and **machine learning**, this project represents a major leap forward in our efforts to answer the question: **Are we alone?** While no signs of alien life have been confirmed yet, the project's discoveries have the potential to reshape our understanding of the **universe**. Whether or not we find **extraterrestrial life**, the advances in technology and international collaboration are already having a profound impact on science and humanity's quest for knowledge.

049 Discovery of Ancient Human Species
Redrawing Our Family Tree

In recent years, one of the most fascinating discoveries in the field of **paleoanthropology** has been the unearthing of previously **unknown** ancient human species. These discoveries have not only broadened our understanding of **human evolution** but have also significantly changed the way we view our own place on the **family tree** of life. The identification of these new human species has sparked fresh debate on how modern humans **evolved** and what our ancestors' lives were like.

One of the most stunning discoveries occurred in 2015 when researchers uncovered the remains of a new human species in a **South African** cave system. Named **Homo naledi**, this species is believed to have lived between 236,000 and 335,000 years ago—overlapping with early modern humans. The excavation of this species revealed over 1,500 fossil elements, making it one of the largest collections of hominid fossils ever found. **Homo naledi** exhibited a mixture of both **primitive** and **modern** traits: a small brain and body size similar to early human ancestors, but with advanced tools and burial practices. This unexpected combination challenged existing theories on the timeline and **complexity** of human evolution.

Just a few years earlier, in 2004, another groundbreaking discovery was made on the **island of Flores**, Indonesia, where researchers unearthed the remains of a **tiny**, ancient human species known as **Homo floresiensis**–nicknamed "the hobbit" due to its small stature. Standing only about 3.5 feet tall, **Homo floresiensis** is believed to have lived as recently as 50,000 years ago, long after the appearance of modern humans. This discovery stunned the scientific community, as it showed that small, **isolated** human populations with unique evolutionary paths existed much later than previously thought. More recently, in 2019, researchers discovered yet another species: **Homo luzonensis**, found in the **Philippines**. This species lived over 50,000 years ago, and like **Homo floresiensis**, it was small in stature. The discovery of **Homo luzonensis** added another layer of complexity to the understanding of how ancient humans migrated and evolved across Asia and the Pacific.

These finds, along with the discovery of **Denisovans** (a previously unknown group of archaic humans identified through **DNA analysis**), are reshaping the narrative of human evolution. No longer is human evolution seen as a **simple, linear progression** from ape-like ancestors to modern **Homo sapiens**. Instead, it is now understood as a tangled web of interactions between different species of early humans, some of which may have even interbred with each other. The discovery of these species demonstrates that **multiple human species** coexisted, and in some cases, interacted with early Homo sapiens.

Applications:
The discovery of these ancient human species is more than just an academic exercise—it has practical applications in several areas:

- **Understanding Human Evolution:** By studying these new species, researchers are gaining new insights into how **modern humans** evolved. The genetic and physical traits of species like **Homo naledi, Homo floresiensis**, and **Denisovans** help scientists piece together the complex **puzzle** of our evolutionary history, shedding light on traits we inherited and why certain species survived while others did not.

- **Forensic Anthropology:** The techniques used to discover and analyze ancient human remains are being applied to forensic science. **Modern forensic anthropology** benefits from advancements in **DNA extraction** and **skeletal analysis**, allowing for more precise identification of human remains in criminal investigations and archaeological excavations.

- **Genetic Research:** The discovery of **Denisovan DNA** in modern humans has provided new information about interbreeding between species. This insight is valuable for geneticists studying inherited traits, diseases, and human resilience. Understanding the **genetic legacy** of these ancient humans could lead to advancements in genetic medicine.

Summary:

The discovery of **ancient human species** such as **Homo naledi**, **Homo floresiensis**, and **Homo luzonensis** has significantly altered our understanding of human evolution. These findings reveal that human evolution was not a **simple, linear** process, but a complex network of interactions between various species. With each new discovery, our **family tree** grows more intricate, offering fresh insights into how we became the species we are today. These revelations not only deepen our knowledge of the past but also provide important contributions to fields like **genetics** and **forensic anthropology**, showing that the story of humanity is far from complete.

050 Plant Communication Discoveries
The Secret Life of Plants

For centuries, plants were seen as **passive organisms**, growing silently, reacting to their environment, but without much interaction. In the past decade, however, a growing body of research has revealed that plants may be **communicating** with each other in ways we never imagined. **Plant communication**—an idea that once seemed confined to science fiction—has become a legitimate and fascinating area of study. What scientists have discovered about the **hidden world of plant interactions** is changing the way we understand nature.

At the heart of this research is the **Wood Wide Web**—a vast underground network of fungi, known as **mycorrhizal networks**, that connects the roots of plants. Through this network, plants can exchange nutrients, send distress signals, and even "warn" neighboring plants of potential dangers, such as herbivore attacks or environmental stressors like drought. This **interconnected system** allows plants to share resources more efficiently and protect one another, painting a picture of the plant world as a cooperative, rather than competitive, ecosystem.

But the revelations don't stop there. Research has shown that plants can release **volatile organic compounds (VOCs)** into the air as a form of **communication**. For example, when a plant is being eaten by herbivores, it can emit **specific VOCs** that "warn" nearby plants to activate their own defenses. Some plants, like acacia trees, release tannins to make their leaves less palatable to herbivores when they sense an attack. Other studies suggest that some plants can even differentiate between the types of herbivores attacking them, adjusting their responses accordingly.

Beyond **chemical communication**, plants also appear to have **electrical signaling systems**. Just like animals, plants exhibit electrical **impulses**, especially when under stress. These impulses, called action potentials, travel through the **plant's vascular system** in response to injury or environmental changes. This electrical activity suggests that plants have a more **dynamic** and **responsive** internal communication system than previously thought.

In addition to communication with each other, plants also engage in **mutualistic relationships** with fungi, bacteria, and even insects. For example, some plants release **chemical signals** to attract beneficial insects, like parasitic wasps, which then attack herbivores threatening the plant. These kinds of **relationships** show that plants not only respond to their environment but actively shape it in ways that promote their **survival**.

Applications:
The discovery of plant communication opens up a new world of possibilities for various industries, from agriculture to conservation. Here are a few key areas where this knowledge is already making an impact:

- **Sustainable Agriculture:** Understanding how plants **communicate** can lead to more sustainable farming practices. By tapping into the natural **defense mechanisms** of plants, farmers could reduce the need for pesticides, instead encouraging plants to communicate and fend off pests naturally. For example, certain crops could be grown next to "companion plants" that enhance pest resistance through the release of **VOCs**.

- **Ecosystem Conservation:** The **Wood Wide Web** underscores the importance of **biodiversity** in maintaining healthy ecosystems. Forests, for example, rely on the complex underground fungal networks that connect trees and plants, facilitating nutrient exchange and enhancing resilience to environmental stress. Conservation efforts are now focusing on **protecting** not just individual species but the networks that sustain entire ecosystems.

- **Crop Improvement:** By studying the signals plants use to warn each other of drought or nutrient deficiencies, scientists are working on ways to **genetically engineer** crops that can better respond to environmental stressors. This could lead to **drought-resistant crops** that adapt faster to changing climate conditions, improving food security in vulnerable regions.

Summary:
Over the past decade, research has revealed that plants are far more **communicative** and **interconnected** than we ever thought possible. Through **mycorrhizal networks, volatile organic compounds,** and **electrical signals**, plants exchange information and even cooperate to defend themselves and share resources. This growing understanding of **plant communication** has profound implications for agriculture, conservation, and crop improvement. As we continue to unravel the **secret life of plants**, we gain new insights into how interconnected and intelligent the natural world truly is.

050+ Deepfake Technology
The New Era of Synthetic Media

Over the past decade, the rise of **deepfake technology** has sparked both fascination and concern across the globe. At its core, **deepfake technology** refers to the use of **artificial intelligence (AI)** to create **highly realistic** but entirely fake videos, images, or audio clips. What sets deepfakes apart from other forms of digital manipulation is the uncanny accuracy with which they can mimic a person's likeness or voice, often making it nearly **impossible** to tell what is **real** and what is **synthetic**.

The technology relies on a subset of **AI** known as **deep learning**, which uses **neural networks** to study and replicate patterns in data. In the case of deepfakes, these networks are **trained** on hours of **video footage**, **photographs**, or audio **recordings** of a person. Over time, the system "**learns**" how to **recreate** that person's face, voice, or movements. It can then superimpose that person's likeness onto another body or create entirely fabricated speech or gestures. The result? A **synthetic version** of a person that can be convincingly placed in any situation.

While the term "**deepfake**" was coined in 2017, the technology had been quietly **developing** for years. Initially, deepfakes were used for relatively benign purposes, such as **special effects** in movies or creating digital avatars. However, as the technology advanced, it became more widely **accessible**, with free **apps** and **software** enabling anyone with basic skills to create deepfakes. This has led to both creative and **malicious uses** of the technology.

On the **creative side**, deepfakes have been used in **entertainment** and **marketing** to bring deceased actors back to life, create engaging visual effects, and personalize advertisements. For instance, the late Carrie Fisher was "**brought back**" as Princess Leia in Star Wars: The Rise of Skywalker using deepfake technology. **Artists** and **filmmakers** are experimenting with the technology to push the boundaries of digital storytelling, offering a glimpse of what the future of media might hold.

However, the **darker side** of deepfakes cannot be ignored. In the wrong hands, deepfake technology has the potential to **cause serious harm**. Fake news, political propaganda, and celebrity impersonations are just a few of the **malicious uses** that have emerged. In some cases, deepfakes have been used to create **compromising videos** of public figures, causing reputational damage or spreading misinformation. The speed at which these falsified videos can go viral only amplifies their impact.

As deepfake **technology** becomes more **advanced**, it is becoming increasingly difficult for the average person to distinguish between **real** and **fake** content. Even experts admit that **identifying deepfakes** is becoming more **challenging**, as the technology continually improves. This has led to widespread concerns about its implications for **trust, privacy,** and **security** in the digital age.

Applications:
Deepfake technology offers both potential benefits and serious risks, depending on how it is used:

- **Entertainment and Film:** In the film industry, deepfake technology allows filmmakers to create **lifelike visual effects, bring back** actors from the past, or de-age performers for specific roles. It opens up new **possibilities** in storytelling and world-building. For example, deepfakes have been used in blockbuster movies to resurrect characters or blend actors seamlessly into CGI environments.

- **Marketing and Advertising:** Brands are experimenting with deepfakes to **personalize ads**. Imagine seeing an ad with a celebrity's face endorsing a product in real-time, tailored to the viewer's preferences. This level of **personalization** could **revolutionize** how companies connect with their target audiences.

- **Political Manipulation and Fake News:** On the darker side, deepfakes are already being used to **manipulate** political discourse. In 2020, a deepfake video of President Obama surfaced, making it appear as if he said things he never actually did. Such examples illustrate how deepfakes can be weaponized to **undermine public trust** in institutions and leaders, potentially affecting elections and international relations.

- **Identity Theft and Cybersecurity:** Deepfakes also pose serious **risks** to **privacy** and **security**. The technology could be used for identity theft, fraud, or blackmail. As it becomes more difficult to distinguish **real** from **fake**, individuals may find their likenesses used without consent, leading to new forms of digital exploitation.

Summary:
Deepfake technology is a **double-edged** sword. On one side, it offers **exciting possibilities** for entertainment, advertising, and creativity, enabling creators to push the boundaries of visual storytelling. On the other side, it presents **serious challenges** in the realm of **misinformation, privacy,** and **security**. As deepfakes become more realistic, **distinguishing** between **real** and **synthetic** content will become even harder, raising urgent questions about how to regulate and mitigate the risks. As society grapples with the implications of this **powerful technology**, the future of synthetic media remains uncertain, but one thing is clear: deepfakes have ushered in a new era of **digital reality**.

THANK YOU NOTE

To my dearest wife,

Thank you from the bottom of my heart for your unwavering faith in me and for all the moments you have allowed me to steal from you in order to write this book.

I know that in writing this book I have missed countless moments with you, moments that I can never get back.

Your support has been a silent but powerful force, helping me to keep going even when work prevented me from giving you the attention you so deserve and need.

Thank you for standing by me, for believing in me and for inspiring me every day to do my best.

I hope I can make up for every moment I have not been by your side these months.

With all my love and gratitude,

O. Bennet

About the author:

O. Bennet

With a degree in **Chemistry** with a specialization in **Polymers**, and additional training in **Computer Systems Administration** and **Programming**, this author brings a blend of scientific and technological knowledge. His passion for **learning** and **discovering** new advances in science and technology fuels his constant **curiosity unleashed** in these fields.

Driven by his love of innovation and progress, he now shares his knowledge with readers through **thought-provoking** publications.

In them, he delves into the most **fascinating marvels of the last decade**, offering accessible and entertaining commentary on the discoveries that are shaping our incredible future.

Through his reading, he opens the door to ideas that will inspire readers to **explore, learn** and **be captivated** by the **ever-evolving** world of science and technology.

Disclaimer of Liability

This book may contain references to trade names, trademarks, service marks, brand names, logos, and other proprietary designations belonging to third parties. These references are included solely for informational and illustrative purposes for the benefit of the reader and do not constitute or imply any endorsement, sponsorship, or affiliation between the author, the publisher, and the owners of such marks.

All trademarks, registered trademarks, product names, company names, and logos mentioned or depicted in this book belong to their respective owners. The use of these names and marks is not intended to infringe the proprietary rights of third parties.

The author and publisher acknowledge and respect all trademarks and copyrights mentioned herein. Mention of specific companies, products or services does not imply any recommendation or endorsement by the author or the publisher, or that these entities have endorsed or approved the book or its contents.

Readers should be aware that the inclusion of any trade name or trademark is for identification and reference purposes only.

www.ingramcontent.com/pod-product-compliance
Lightning Source LLC
Chambersburg PA
CBHW071456220526
45472CB00003B/822